T0331226

Migration, Community and Identity

Migration, Community and Identity analyses experiences of migration to rural Wales from 1965 to 1980. It focuses on people who were part of the era's counterculture, looking for an escape from mainstream society. Using original interviews, the book shows why people moved and how the move shaped their lives and identities. Drawing together geographical and historical research, this book explores the significance of this migration phenomenon. It provides a unique insight into late-twentieth-century Welsh society and shines a new light on the counterculture itself. Through analysing the experience of life in Wales and ongoing developments in the migrants' sense of identity, it argues that rather than being a uniform group, the counterculture encompassed a diverse range of beliefs and aspirations.

The book will be suitable for upper-level undergraduates and above. The broad range of themes covered in this book is relevant not only to rural and historical geographers and migration researchers but also to those interested in sociology, anthropology, and the modern history of Britain and Wales. The theories and concepts discussed have global appeal and will be of interest to those studying similar migration phenomena elsewhere.

Flossie Caerwynt is a Research Associate in the Department of Geography and Earth Sciences at Aberystwyth University. She has an interest in heritage studies and the impact of the past on the present, with a particular focus on alternative lifestyles and experiments in reimaging society.

Routledge Research in Culture, Space and Identity

Series Editor: Peter Merriman

The *Routledge Research in Culture, Space and Identity Series* offers a forum for original and innovative research within cultural geography and connected fields. Titles within the series are empirically and theoretically informed and explore a range of dynamic and captivating topics. This series provides a forum for cutting edge research and new theoretical perspectives that reflect the wealth of research currently being undertaken. This series is aimed at upper-level undergraduates, research students and academics, appealing to geographers as well as the broader social sciences, arts and humanities.

Creative Engagements with Ecologies of Place
Geopoetics, Deep Mapping and Slow Residencies
Mary Modeen and Iain Biggs

Comics as a Research Practice
Drawing Narrative Geographies Beyond the Frame
Giada Peterle

Spaces of Puppets in Popular Culture
Grotesque Geographies of the Borderscape
Janet Banfield

Migration, Community and Identity
Countercultural Lifestyle Migration to Rural Wales, 1965–1980
Flossie Caerwynt

Spatialities of Speculative Fiction
Gwilym Lucas Eades

For more information about this series, please visit: www.routledge.com/Routledge-Research-in-Culture-Space-and-Identity/book-series/CSI

Migration, Community and Identity

Countercultural Lifestyle Migration to Rural Wales, 1965–1980

Flossie Caerwynt

LONDON AND NEW YORK

First published 2024
by Routledge
4 Park Square, Milton Park, Abingdon, Oxon OX14 4RN

and by Routledge
605 Third Avenue, New York, NY 10158

Routledge is an imprint of the Taylor & Francis Group, an informa business

British Library Cataloguing-in-Publication Data
A catalogue record for this book is available from the British Library

Library of Congress Cataloging-in-Publication Data
Names: Caerwynt, Flossie, author.
Title: Migration, community and identity : countercultural lifestyle migration to rural Wales, 1965–1980 / Flossie Caerwynt.
Description: Abingdon, Oxon ; New York, NY : Routledge, [2024] | Series: Routledge research in culture, space and identity | Includes bibliographical references and index.
Identifiers: LCCN 2023025259 (print) | LCCN 2023025260 (ebook) | ISBN 9781032415529 (hardback) | ISBN 9781032415536 (paperback) | ISBN 9781003358671 (ebook)
Subjects: LCSH: Migration, Internal—Wales—History. | Group identity—Wales—History. | Counterculture—Wales—History.
Classification: LCC HB2255 .C347 2024 (print) | LCC HB2255 (ebook) | DDC 304.8/0409429—dc23/eng/20230801
LC record available at https://lccn.loc.gov/2023025259
LC ebook record available at https://lccn.loc.gov/2023025260

ISBN: 978-1-032-41552-9 (hbk)
ISBN: 978-1-032-41553-6 (pbk)
ISBN: 978-1-003-35867-1 (ebk)

DOI: 10.4324/9781003358671

Typeset in Times New Roman
by Apex CoVantage, LLC

For my boys.

Contents

Figures

Tables

Acknowledgements

This book is the result of my PhD research into countercultural migration, and as such, the majority of the thanks must go to the people who supported me through this original research. This encompasses so many people – from the tireless National Health Service (NHS) staff who got me through a mid-project hip replacement to the many kind helpers who brought me food during the 2020 lockdown – that it is impossible to list them all individually. Special credit must of course go to my PhD supervisors, Pete and Gareth, for their invaluable support and advice, and to all the staff at the Wales Institute of Social and Economic Research and Data (WISERD), who supported (and put up with!) me taking time to work on this rendition of the original research. Thanks also go to the friends who did PhDs alongside me for the countless much-needed Arts Centre lunches and evening drinks, particularly Anna and Keziah, for being the very best of best friends.

Outside of academia, I couldn't have done the research at all if it hadn't been for the very best Mum and Dad in the world. Thanks also have to go to Cal, of course, for being so committed to the whole thing that he married me halfway through. A very special mention must also go to Falstaff and Feldspar, the very best little kittens, who have always been keen to add their own thoughts to the manuscript. Sadly, their contributions may not have made the final draft.

Introduction

> Society was going in a very material direction at those times and we didn't really fancy it.
>
> – George

> We wanted to opt out of city life . . . we wanted to be able to grow our own food and have the freedom of living in the country.
>
> – Chloe

> I sort of try to be what I believe in.
>
> – Barbara

All these quotations come from people who moved to Wales as part of the counter-culture movement of the 1960s and 1970s. The stories they tell are imbued with a sense of determination, a firm belief in a different way of doing things, and, at the same time, a feeling of making their way in a world which was not quite familiar to them. The 1970s had been marred by political and economic turmoil, as the oil crisis and miners' strikes led to the adoption of the three-day working week, increasing Irish Republican Army (IRA) violence threatened people's safety, and the threat of nuclear war lay on the horizon (Grindrod 2013: 363; Weart 2012: 204). It is hardly surprising that people like George, Chloe, and Barbara wanted to escape this chaotic and unpredictable world and find somewhere to build a better life for themselves. For them, and for the many others like them, moving to Wales was the solution they had been looking for.

In this book, I trace the lives and experiences of people who moved to Wales as part of this reaction against the events of the 1960s and 1970s – the 'counterculture' as it came to be known. This counterculture created the Temporary Tutonomous Zone described by Hakim Bey 'a space which enabled the fleeting suspension of usual rules and mores' (Whiteley 2010: 50). In turn, this creation of a new kind of space in response to wider events is reflected in the lives of those who moved to rural Wales. The combination of rural and countercultural influences means that

DOI: 10.4324/9781003358671-1

aspects of the migrants' stories resemble the 'places on the margin' described by Rob Shields, creating 'near sacred *liminal zones* of otherness' (1991: 5–6, emphasis in original). However, these migrants' stories have received little recognition in academic literature to date. Research into the counterculture more widely tends to focus on urban areas (Gilbert 2006; Nelson 1989; Rycroft 2002, 2003, 2007, 2011). There is also a growing body of work on festivals and other social phenomena which accompanied the counterculture, such as protest movements or New Age travellers (Bennett et al. (eds.) 2016; McKay 1996; Jones and O'Donnell (eds.) 2010; Pile and Keith (eds.) 1997; Sharp et al. (eds.) 2000). When studies of rural alternative lifestyles in Britain do exist, they often focus on communes or other forms of alternative community rather than on individuals (Pepper 1991; Rigby 1974a, 1974b). Or else they cover a different time period, focusing, for instance, on Victorian and early-twentieth-century pastoralism or the interwar enthusiasm for going back-to-the-land in Wales (Gruffudd 1994; Marsh 1982; Matless 2016). Keith Halfacree's overview of back to the land experiments since the 1960s is one of the few to deal with the topic, and he acknowledges that 'geographers have poorly researched this strand, especially in contrast to an extensive body of work on more conventional forms of counter-urbanisation, for example' (2006: 309).

This book shares these neglected stories – the stories of individuals who moved alone or in family units and who were part of a rural counterculture that was at the same time similar to and different from the urban version. While urban countercultures, festivals, communes, and pastoralism are all undoubtedly significant, they do not cover the whole story of alternative living in the 1960s and 1970s. People were involved on an individual level across the country – not just in urban centres – paying attention to what the counterculture was about and incorporating its ideas into their lives in the ways that they could. The experiences of those who moved to rural Wales represent a different side of the counterculture from that which has been discussed before. They are an important example of the rural branch of the counterculture and of the ways in which the counterculture was intrinsically connected to the wider social and political changes happening across the country. Theirs is a common story from the time, but one which has not been well told before.

In sharing and analysing the stories told by contributors to this project, I hope to find the answers to questions grouped around three research themes, each of which forms one part of this book. The first theme is their motivations. I want to know why so many people moved to rural Wales during the late 1960s and 1970s – what drove them to leave their previous homes and what made them choose Wales in particular? Existing research can tell us how the counterculture came into being as an idea, but it does not explain what attracted people to it on an individual level – whether it was a conscious choice, for instance, or if people just became swept up in it because of decisions they were making anyway. The second theme is practices. This theme's purpose is to find out what kind of alternative knowledge and beliefs these migrants brought to rural Wales, and more specifically, what was it like putting these into practice on an everyday level? I do not just want to know about the theory behind these lifestyles, I also want to know about the reality of living these lives. This is another angle which is missing from existing narratives of the counterculture movement. It is all very well knowing about big ideas and overarching themes, but for the individuals involved, this was something they experienced in the moment, and I want to know what those moments were like. And finally, belonging. How did these migrants fit in with existing communities? What impact did living in rural Wales have on their identities and those of their descendants? These migrants and their families are still very much part of Welsh communities today; this is not a story consigned to the past but one which is still unfolding around us. And it is important to remember that the counterculture did not happen in isolation; while these migrants may have been looking for an alternative to the lives they were living, it was impossible to escape those lives entirely.

In order to uncover the answers to these questions, I based this research on oral history interviews with 55 people who moved to Wales between 1965 and 1980. It is the stories gathered through this process that have shaped the research more than anything else. These stories act as a medium for people to express the connections between their lifestyle, identity, and the place they have chosen to live. Anthropologist Keith H. Basso describes:

> Whenever the members of a community speak about their landscape – whenever they name it, or classify it, or evaluate it, or move to tell stories about it – they unthinkingly represent it in ways that are compatible with shared understandings of how, in the fullest sense, they know themselves to occupy it.
>
> (1988: 101)

Place, lifestyle, and identity are inextricably connected through story, and it is this intertwining which the project grows out of. There are, of course, limitations to this approach. I have not included countercultural migrants to other parts of the UK, for instance, so I do not know how their experiences may have varied or what that might mean for the counterculture as a whole. In order to keep the project to a manageable size, this is a necessary restriction, but a study of other areas could be an interesting avenue for further research. Another limitation is that the vast majority of the contributors to this project were people who still live in Wales. It is very likely that more people will have moved to Wales as part of the counterculture but later given up and gone elsewhere. The stories covered here are ones of making a success of the move – the voices of those who found the experience too difficult are not included. This is primarily due to the difficulties of finding and interviewing people no longer resident in the area – even with the internet, it is far easier to recruit participants within a defined geographical area. The other reason for this omission is, as mentioned earlier, the need to keep the sample to a manageable size. However, it is still a limitation that is worth remembering and, again, one which presents a possibility for further research.

There is a strong gender bias in the interviewee data: 39 women contributed, but only 15 men. There are a number of potential reasons for this, ranging from the recruitment techniques (which emphasise social media and are more likely to be used by women than men) to gender-based differences in attitudes towards nostalgia and the past (for a variety of cultural reasons, women in the Global North are more likely to develop stronger autobiographical memories) (Newman et al. 2019: 19; Jackson 2012: 1002; McLean et al. 2007: 265–266; Fivush 2011: 569). Regardless of why there is a gender bias, it is good to keep these differences in mind. They may well have had an impact on the data itself, and running the project again with a primarily male cohort would likely uncover different information. That said, it is important not to assign too much importance to a single aspect of the migrants' identities. Victoria Land and Celia Kitzinger argue that:

> Any given individual can be characterised by a wide range of category terms taken from many different category sets, including, for example, gender, sexuality, political alignment, ethnicity, age, nationality, religion, occupation, place of residence, health status, family position and so on. . . . The speaker who is a 'woman' is also, for example, a lesbian, a Pagan, an environmentalist, a diabetic, a sister and so on.
>
> (2011: 48)

The intersectionality of identity is crucial here. It is impossible to assign any feature of the contributors' stories to a single characteristic or indeed to any particular combination of characteristics. They all combine to create the unique way each individual remembers and recounts their past.

Oral history is, of course, not without its flaws. Memory is subject to distortion, and it can be difficult to tell how accurate people's recollections are. Then again, Schudson points out that 'the notion that memory can be "distorted" assumes that there is a standard by which we can judge or measure what a veridical memory must be' (1995: 346). This idea of a single, verifiable truth is something that is the subject of much controversy in the humanities. I argue that, regardless of whether a single true version of any event exists, memories (and, indeed, the ways they are distorted) can still reveal useful information. Alistair Thomson sums up the problems and benefits of oral history well when he states that:

> At one extreme personal testimony is criticised as a flawed and unreliable historical source because of bias, self-justification or forgetfulness. . . . At the other extreme, enthusiasts and especially popular historians respond that personal narratives illuminate the lived experience and meanings of historical events, and lives of groups of people – so called 'ordinary' men and women – who are unlikely to be documented in the archives of the governing classes.
>
> (2011: 101)

In short, there are advantages and disadvantages to using oral history. It excludes some stories but encourages others. From its inception, it has focused on the voices of ordinary people (Ashplant and Smyth 2001: 3; Abrams 2016: 4; Samuel 2012: 191). These are the stories that I set out to hear, and they are the stories I was able to uncover. They are messy, subjective, and flawed, but in being so, they tell us more than we could hope to know about the hopes, dreams, experiences, and perceptions of the people telling them.

In truth, the idea for this research came about in a rather roundabout way. I knew right from the outset that I wanted to do research using oral history, and I knew I wanted that research to be based in Wales. I knew that it would have something to do with rural areas and alternative ways of being and seeing things. But I had not yet worked out how to bring all of these elements together. The lives of countercultural migrants to Wales provided the answer. The counterculture movement was, after all, focused on finding alternatives to what it saw as the flawed parts of modern life; it was about looking at things differently and reimagining society in light of this. As I mentioned at the beginning, a big part of this centred around moving to the countryside, and Wales was a prime location for this particular branch of lifestyle migration. And since many of the people who came to Wales as part of this movement are still living here, oral history is the obvious way of finding out their stories. Not only that, but I found that these people were very keen to share their experiences to help capture this unique movement before it was forgotten. This book shares their stories, but in doing so, it achieves more than simply recounting their experiences. It also provides a new analysis of a key cultural movement, hopefully helping to shape the way that the counterculture is conceptualised.

While I may have initially been attracted to this project because of its potential to incorporate the elements I wanted, as I journeyed through the research, I found more and more similarities appearing between my own experiences and those of the contributors to the project. My grandparents moved to Wales in order to keep a small flock of sheep after retiring in the early 1990s – too late to be part of the exact movement discussed here but very much inspired by the same ideas. And since I grew up visiting them, the narrative was well ingrained into my own life too – so much so that I would make the move to Wales myself many years later (though I have yet to acquire any sheep). Other aspects of the countercultural migrants' stories resonated with me as well: I have been vegetarian since long before the current trend for giving up animal products; I have been involved in community-supported agriculture programmes; and I have worked to promote environmental awareness and sustainability. I know the impact that moving countries has on identity, and the challenges of fitting into a new community. The stories shared here are not my own, and I make no claim to them. Nor did I choose the project because of these connections. But the connections between these stories and my own life have nevertheless helped to shape this research and have given me an insight into the enormous impact that the counterculture movement had – so much so that I can feel its echoes in my own life, more than 50 years later.

It's not only my life that contains parallels to the stories of the countercultural migrants. Perhaps the most significant reason for pursuing this project is because of the rapidly mounting similarities between the events that pushed the counterculture into existence and the events that occurred while I carried out this research. Take environmentalism, for instance. In December 1972, the crew of Apollo 17 took the first full image of Earth from space. This was the first time we had come face to face with the full extent of our world, its fragility, and its finite limits (Dryzek 2005). The release of this image had a profound impact on people's perceptions of the planet and is credited by many as being one of the key events that triggered the start of the environmental movement. The image was seen to represent the interconnectedness of life on Earth, described by Denis Cosgrove as the 'Whole-Earth' reading of the image (1994). As one contemporary observer later recalled:

> Like everyone else, I had seen many artists' depictions and models of the Earth, but none of those had the same visceral impact on me as did seeing our planet surrounded by the blackness of space, sitting there alone with nothing to support it. . . . Seeing for the first time 'our' 'Blue Marble' in a vast void has brought home to many how much we depend on the Earth and that it is our responsibility to protect the health and wellbeing of this collective, interdependent eco-system.
>
> (Wuebbles 2012)

Scientists estimate that the rapid rate of environmental change that we are now experiencing began sometime around the 1950s (Goudie 2013: 329). The rapidity of this change meant that in the 1960s and 1970s, people's attention was being drawn to the environment in new ways (Sheail 2002: 1–2). They were faced with the speed of change on one hand and the Blue Marble's reminder of Earth's finite resources on the other – an image both comforting in its reminder of unity and terrifying in its limits (Cosgrove 2001: 263). So people started to engage more with campaigns for reducing pollution and waste and finding ways to live lightly on the planet. Friends of the Earth made history by taking the first piece of direct action in British environmentalism when, in 1971, they dumped glass bottles outside the Schweppes office in London to protest against them being un-reusable (Doherty et al. 2000: 5).

If a similar protest were to be reported on the news tomorrow, it would not feel remotely out of place. A narrative of rising awareness of our environmental impact and increasingly heated attempts to make something happen to combat it is familiar to us today. In the four and a half years I spent writing this project, the world witnessed major wildfires on every continent bar Antarctica. A climate emergency was declared in 1,799 jurisdictions, encompassing parts of 31 countries in total (Climate Emergency Declaration 2020). Extinction Rebellion's statement of open rebellion against the UK government led to the formation of hundreds more Extinction Rebellion groups around the world, with an official International Rebellion beginning in 2019 (Knights 2019: 10). The same issues around waste and environmental impact that led Friends of the Earth to dump Schweppes bottles are once again driving protests, this time on a global scale.

And there were other parallels too – plenty of them. Britain's relationship with the European Union is another big one; in 1975, Britain held a referendum over whether to continue with the EU membership that had only been gained two years previously (Evans 2018: 127–128). In the subsequent 2016 referendum, right at the start of this project, Britain voted to leave the EU. The negotiations which followed have been a constant background to this research, including the interview collection process. Then there is the question of race relations, as the 1960s and 1970s are remembered as the era of Commonwealth immigration, civil rights, and the 'birth' of multiculturalism. It was in the midst of this that Enoch Powell made his infamous 1968 'Rivers of Blood' speech – now regarded as 'one of the most racist speeches in British history' – as part of an effort to curb immigration (Eddo-Lodge 2017: 117). And in 2018, I watched as the Windrush scandal revealed how the same immigrants that Powell had railed against were treated as illegal entrants decades later (Hewitt 2020: 108–109). In 2020, race equality was thrown into focus yet again as the murder of George Floyd by a Minneapolis police officer set off a chain of Black Lives Matter protests that spread across the globe and mirrored the civil rights protests of the 1960s and 1970s to a frightening extent (Joseph-Salisbury 2020). Likewise, feminism is still very much on the agenda. In 1968, female machinists at the Ford factory in Dagenham went on strike against unequal pay, the start of a flurry of similar protests that would happen around the country over the following years (Nicholson 2019: 321–322). In 2017, following the #MeToo movement and the election of Donald Trump as President of the US, more than 100,000 people participated in a Women's March in London, protesting against continuing inequalities (Emejulu 2018: 268). We may have moved from the second wave to the third wave, but feminist issues are still very much present.

There is also a growing dislike of urbanisation and mistrust of city living. In 1968, Ronan Point, a newly built 22-storey tower block in East London, partially collapsed due to a gas explosion on an upper floor, killing four people and injuring 17. It led to a nationwide scare over the safety of such dwellings and permanent changes in building regulations (Grindrod 2013: 330–333; Pearson and Delatte 2005: 172). Fears over the safety of high-density housing and the wider inequalities surrounding it returned when Grenfell Tower in West London caught fire in 2017. Leaving 72 people dead and a further 70 injured, this was the UK's worst peacetime fire since the Victorian Era. Like Ronan Point, it led to a renewed distrust of urban high-rises and ongoing investigations into building regulations (MacLeod 2018: 460). Perhaps it is not surprising then that, as the coronavirus pandemic adds to the growing number of anxieties present in modern life, people are once again turning to the idea of a new life in the countryside as a solution. Enquiries about rural properties increased by 126% in 2020, as more and more people sought what they perceived to be a better life (Jones 2020; BBC News 2020). Perhaps we're about to see a new wave of countercultural migrants; the conditions certainly seem right.

Some of these events are more significant than others to the stories that will be shared here, but they are all pertinent to understanding the world that this group of migrants were living in and still live in. I am not saying that everything happening today is the same as what happened in the 1960s and 1970s, but I am saying that the similarities between the two add an extra layer of poignancy and significance to the research. The world these migrants inhabited at the start of their journeys is not so different from our own; their stories hold a sense of familiarity. And because these migrants and their families continue to live here in Wales, their stories are ongoing, encompassing both sides of the parallels.

This book is organised into three parts, each of which deals with one of the research themes: motivations, practices, and belonging. Each of these contains three chapters, each dealing with a different aspect of the migrants' experiences. In Part 1, we begin the migrants' story, appropriately enough, at the start of their journey to Wales. Over the course of three chapters, we will meet the core cast of recurring characters, learning precisely what drove each of them to seek out a new life in the countryside. Through stories from Lesley, Barbara, Josie, Judith, and Ian, we will see just how diverse the backgrounds of these migrants were. We will learn how their experiences were connected to what was going on at the time, both in political and economic terms and on a social level. Then we will turn our attention to why and how they chose to come to Wales in particular, and in so doing, we will be able to find the answer to our first research question: understanding why people moved to rural Wales.

In Part 2, we learn about the day-to-day practicalities of being 'alternative'. This begins with the challenges faced by those migrants who attempted to adopt a self-sufficient lifestyle, with its harsh learning curve and history steeped in gender stereotypes. But since self-sufficiency was not for everyone, we will also look at the ways these migrants earned money, from the jobs they took on to the businesses they started themselves. We then turn to what they did away from work: the hobbies they had, the things they read, and their spiritual beliefs and practices. To do so, we will be drawing together literature about alternative lifestyles and the ideas and knowledge associated with them, alongside additional historical sources. Altogether, this should lead us to the answer to the second research question, so that we understand what kind of knowledge and beliefs they brought with them and what it was like putting them into practice.

Part 3 addresses the way people fit into life in their new Welsh communities, how they formed relationships, and how they found space for themselves. This is where we learn about the anti-English tensions that were brewing in Wales at the time, about the role played by the Welsh language, and, perhaps most importantly, about the way the move to Wales impacted the migrants' sense of identity. In these chapters, the interview data is analysed in the context of theories of national and cultural identity and migration. This will provide us with the answer to the questions relating to belonging: how did these migrants fit in with existing communities, and what impact did living in rural Wales have on their identities and those of their descendants?

Before we delve into these stories, it will be useful for us to take a moment to consider exactly what we mean by 'counterculture'. It's the central theme of this book, the thing which unites all of the stories told. But what precisely *is* counterculture? As early as 1960, American sociologist J. Milton Yinger came up with the term 'contraculture' to refer to 'the creation of a series of inverse or counter values (opposed to those of the surrounding society) in face of serious frustration or conflict' (1960: 627). Theodore Roszak's 1969 volume, *The Making of a Counter Culture*, was influential in spreading counterculture as a concept in its own right (Partridge 2018: 1). The trouble is that everyone seems to have had a different opinion on precisely how counterculture should be defined, and these opinions often differ substantially. Simon Rycroft highlights the difficulty of defining the counterculture in his research into 'Swinging London', stating that 'boundaries between hegemonic and counter-hegemonic cultures are easily blurred' (2002: 567). But boundaries of some kind must still be drawn, at least to start with, if the term is to be used meaningfully. Elizabeth Nelson began her account of British counterculture by asking a similar set of questions about its definition:

> What was the counter-culture? How has it been defined? What did the term mean to those who were a part of it? These questions pose a further one: what was the culture which the counterculture sought to counter?
>
> (1989: 1)

For Nelson, counterculture was a 'struggle against the dominant culture' which happened from 1966 to 1973 and was primarily an urban movement (ibid: 4 & 108). Several characteristics of this definition appear in other places. Rycroft, for instance, also treats the counterculture as primarily urban. It is common to define a set timeframe for the counterculture, which usually runs (as Nelson's does) from the mid-1960s to the early-1970s (Rycroft 2002; Gair 2007: 9–10). These definitions interpret 'counterculture' in a very narrow way. Others have taken a broader view. The Oxford English Dictionary defines counterculture as 'a radical culture, esp. amongst the young, that rejects established social values and practices; a mode of life opposed to the conventional or dominant' (Oxford English Dictionary 2020a). This is similar to the definition used by Yinger in his 1982 book on counterculture: 'A set of norms and values of a group that sharply contradict the dominant norms and values of the society of which that group is part' (3). This is the version of counterculture that underpins this book. Focusing on overarching values that go against convention is a model that can be applied to any time or place, meaning it can be applied to all of the migrants at any point in their lives. This allows us to follow their stories through their time in Wales through the lens of counterculture; it becomes both a cultural phenomenon that led to their migration and a way of interpreting their stories.

Britain is often regarded as the centre point of the counterculture, but it was far from being a unique British phenomenon. Many would argue that the US was an equally important influence, as the famous summer of 1967 was a defining moment in the creation of 'hippies' as a distinct cultural group. Over the course of the decade, thousands of communes were established across the US (Hall 1968: 1; Miller 1999: xiii–xiv). As in the UK, the American counterculture was closely linked with protest movements, including the fight for civil rights and anti-Vietnam War activism. Several significant events in the early history of environmental counterculture originated in the US, including the first Earth Day and the publication of Rachel Carson's *Silent Spring* (McKay 2005: 49–50). Such was the influence of the US on the countercultural scene that it has been subjected to a great deal of academic interest and scrutiny (see, for example, Kersen 2021; Bach 2020; Miller 2011 [1991]; Kirk 2007).

Although studies of counterculture frequently focus on Britain and/or the US, countercultures developed in many countries across the world, sometimes prompted by British or American influence and sometimes independently. Countercultural movements developed across Latin America, linked to transnational exchanges of culture and simultaneous revolutionary movements (Dunn 2014: 432–433, 2016: 8). There was a thriving counterculture in Ukraine during the 1970s, as well as in Czechoslovakia, Yugoslavia, Poland, and Russia, though in these countries young people rebelled against Communist Party control rather than capitalism, as in the West (Risch 2005: 565–566; Yoffe 2020: 112–113). France, Germany, Spain, Australia, and India all developed their own countercultures, each sharing similarities while also maintaining a unique character (Brown 2013: 817; Lutfi 2015: 42; Newton 1988: 53; Warne 2007: 309–310; Martinez 2007).

While the concept of counterculture may be relatively new, the idea of studying subcultures dates back to at least the 1940s (Gelder 2005: 1). Extensive studies of subcultures emerged in the UK as part of the work done by the Birmingham Centre for Contemporary Cultural Studies (CCCS), established in 1964. Three features dominated this early body of work. Firstly, as demonstrated in the influential work by Hall and Jefferson (eds.) (2006), Cohen (1997), and Willis (1977), among others, it tended to understand subcultures as primarily a youth phenomenon. This excluded any potential to identify subcultural behaviour in adults and contributed to the narrative of young people's culture shifting away from the culture of their parents and grandparents. Secondly, it tended to view subcultures through the prism of class rather than other factors such as gender or ethnicity. Although there were researchers prioritising other characteristics, such as Angela McRobbie's work on gender (2000) or Dick Hebdige's focus on ethnicity (1979), they tended to be positioned outside the main bulk of subcultural research. McRobbie, in particular, has written about the flaw in this dominant focus, pointing out that while class was the overriding concern of most cultural research at the time, it was a primarily theoretical concern; on the ground, politics played out according to gender, race, and age more so than class (2000: 2–3). Finally, the CCCS and associated work focused on subcultures as forms of leisure rather than all-encompassing lifestyles. This final shortcoming permeates the majority of counter/subculture research from this era and contributes to the stereotype of the leisure-dominated hippy lifestyle by erasing their working lives from the dominant narrative.

Later research into subcultures, after the initial outpourings of the CCCS, continued to be influenced by the echoes of the CCCS's impact. The convention of the CCCS (and other research at the time) was not to focus on the regional distinctiveness of the subcultures being studied (McRobbie 2000: 1) and eventually this tendency morphed into the trend for countercultural and subcultural researchers to focus overwhelmingly (and generally uncritically) on urban cultures. As mentioned earlier, Elizabeth Nelson's book on the underground press focuses almost solely on urban counterculture, arguing that the counterculture is primarily an urban movement (1989: 108). When rural aspects of the counterculture are mentioned, it is only in passing. The one big exception to this is her analysis of *Communes*, a journal explicitly targeted at the rural counterculture that Nelson argued was peripheral to counterculture proper (ibid: 130). This example demonstrates the difficulty of pinning down parameters for the counterculture – no sooner is one parameter defined (for example, 'urban') than something comes along which contradicts it (for example, *Communes*). Simon Rycroft's work also highlights the difficulty of defining the counterculture in his research into 'Swinging London', arguing that 'boundaries between hegemonic and counter-hegemonic cultures are easily blurred' (2002: 567). This blurring of boundaries between cultures is central to the stories shared with me throughout this research project. Focusing on the urban side of the counterculture gives a partial view, one dominated by the underground press, art, media, and shifting consumer cultures.

More important to this study is the rural branch of the counterculture. It is impossible to draw a firm boundary between the two; they both share some characteristics, such as the reimagination of relationships between nature and technology, but diverge on other points. Rural counterculture comes in a variety of forms, the best known of which is the back-to-the-land movement of the 1960s and 1970s, which many (though not all) of the contributors to this research were influenced by. This book moves away from the narratives which have dominated countercultural and subcultural research, focusing instead on something which rarely receives attention: the day-to-day lives of rural countercultural migrants. The focus on the all-encompassing experience of the everyday means that this research owes just as much (if not more) to research into everyday geographies as it does to counter/subculture research. Geography's interest in the everyday arose as part of the wider 'cultural turn', and so it is not only geographers who have contributed to the ideas presented here. Psychologists, sociologists, anthropologists, and so on have all influenced the ideas of 'the everyday' as a topic of study, and all of these influences have fed into this book in one way or another (Highmore 2002). In its focus on the impact that the shift to a countercultural lifestyle has on the everyday rhythms of people's lives, this book was indirectly influenced by the ideas of time-geography first developed by Torsten Hägerstrand. In particular, it was impacted by work which looks at the way that bigger changes in social structures and institutions affect people's day-to-day lives, such as that by Pred (1981) or Thrift (1981, 1983, 1984). According to time-geography,

> the details of everyday life, – the intersections of individual path and institutional project as well as the individual execution of extra-institutional projects, – at one and the same time are rooted in past everyday details and serve as the roots of future everyday details.
>
> (Pred 1981: 5 & 10)

In other words, the institutions and structures which define our everyday activities are themselves defined by our everyday activities.

The term 'hippy' often comes up in literature about counterculture, particularly the back-to-the-land movement. Contributors to this project sometimes chose to use 'hippy' to define themselves. However, the colloquial term 'hippy' has become so loaded with stereotypes that it does not always accurately reflect the movements it is used to describe. A quick online image search for 'hippy' brings up a plethora of images of people dressed in what are supposedly 'hippy' costumes: tie-dyed in bright psychedelic colours, with lots of jewellery, big sunglasses, long hair, and splendidly impressive 1970s-esque false moustaches. This is not a helpful image and is one that needs to be dispelled if we are to properly understand the countercultural migrants' experiences. An alternative understanding of 'hippy' can be found in John McCleary's book *Hippie Dictionary*, which states that a hippy is:

> A member of a counterculture that began to appear in the early 1960s and expressed a moral rejection of the established society. . . . The true hippie believes in and works for truth, generosity, peace, love, and tolerance. The messengers of sanity in a world filled with greed, intolerance, and war.
>
> (2004: 246)

This is a much more positive view, but even it is overly narrow. A radically different interpretation comes from history writer Christopher Bray, who argues that:

> The hippies were genuine conservatives disgusted at what they saw as empty-headed materialism (in both the consumerist and secularist senses of the word) of an increasingly irreligious culture. At heart, the hippies were missionaries without a church. Far from being the beginning of something, then, the counterculture was the end of something – a world view that had emphasised not the here and now of consumer capitalism but an afterlife that was yours to work for.
>
> (2014: xix)

There are echoes of accuracy in this – reactions against consumerism will certainly play a big role in the stories that follow – but I had argue against the idea that hippies were essentially conservatives. What these definitions give us is an idea of the kinds of images and narratives that have been plastered over the lives of those who moved to Wales. It is something to be aware of and bear in mind as we approach the task of interpreting the lives of those who were part of the counterculture.

The stories shared and told throughout this book are significant for a wide range of reasons. They are significant to me personally because of the unexpected way they resonated with my own life. But they are also significant because of the way they simultaneously parallel and are part of the political, social, and economic events that are happening now. My goal is not to make a comparison or to argue that we should be following the counterculture's ideas for living today. But it is nevertheless valuable to reflect, at the start of our journey through the counterculture, on just how surprisingly relevant it is. Despite this, it has been surrounded by stereotypes and arguments over definition and scope. Perhaps for these reasons, or for others, this is a story that has not been properly told before. But it is one which the people who experienced it are eager to share. This is our chance to listen to their stories, to understand what happened and why, and to see how this movement of people shaped Wales as we know it today.

References

Abrams, Lynn (2016) *Oral History Theory*. Abingdon: Routledge.

Ashplant, T. G. and Smyth, Gerry (2001) 'Schools, Methods, Disciplines, Influences'. In Ashplant, T. G. and Smyth, Gerry (eds.) *Explorations in Cultural History*. London: Pluto Press.

Bach, Damon R. (2020) *The American Counterculture: A History of Hippies and Cultural Dissidents*. Lawrence, KS: University Press of Kansas.

Basso, Keith H. (1988) '"Speaking with Names": Language and Landscape among the Western Apache'. *Cultural Anthropology*, 3(2): 99–130.

BBC News (2020) 'Coronavirus: "I Hated My Flat during Lockdown"'. Available at: www.bbc.co.uk/news/business-53670199. Accessed 21 October 2020.

Bennett, A., Taylor, J. and Woodward, I. (eds.) (2016) *The Festivalisation of Culture*. Abingdon; New York: Routledge.

Bray, Christopher (2014) *1965: The Year Modern Britain Was Born*. London: Simon & Schuster.

Brown, Timothy Scott (2013) 'The Sixties in the City: Avant-Gardes and Urban Rebels in New York, London, and West Berlin'. *Journal of Social History*, 46(4): 817–842.

Climate Emergency Declaration (2020) 'Climate Emergency Declarations in 1799 Jurisdictions and Local Governments Cover 820 Million Citizens'. Available at: https://climateemergencydeclaration.org/climate-emergency-declarations-cover-15-millioncitizens/. Accessed 23 October 2020.

Cohen, Phil (1997) *Rethinking the Youth Question: Education, Labour and Cultural Studies*. Basingstoke: Macmillan.

Cosgrove, Denis (1994) 'Contested Global Visions: One-World, Whole-Earth, and the Apollo Space Photographs'. *Annals of the Association of American Geographers*, 84(2): 270–294.

Cosgrove, Denis (2001) *Apollo's Eye: A Cartographic Genealogy of the Earth in the Western Imagination*. Baltimore: The John Hopkins University Press.

Doherty, B., Paterson, M. and Sees, B. (2000) 'Direct Action in British Environmentalism'. In Doherty, B., Paterson, M. and Sees, B. (eds.) *Direct Action in British Environmentalism*. London; New York: Routledge.

Dryzek, J. S. (2005) *The Politics of the Earth: Environmental Discourses*. Oxford: Oxford University Press.

Dunn, Christopher (2014) '*Desbunde* and Its Discontents: Counterculture and Authoritarian Modernization in Brazil, 1968–1974'. *The Americas*, 70(3): 429–458.

Dunn, Christopher (2016) *Contracultura: Alternative Arts and Social Transformation in Authoritarian Brazil*. Chapel Hill, NC: The University of North Carolina Press.

Eddo-Lodge, Reni (2017) *Why I'm No Longer Talking to White People about Race*. London: Bloomsbury.

Emejulu, Akwugo (2018) 'On the Problems and Possibilities of Feminist Solidarity: The Women's March One Year On'. *IPPR Progressive Review*, 24(4): 267–273.

Evans, Adam (2018) 'Planning for Brexit: The Case of the 1975 Referendum'. *The Political Quarterly*, 89(1): 127–133.

Fivush, Robyn (2011) 'The Development of Autobiographical Memory'. *Annual Review of Psychology*, 62: 559–582.

Gair, Christopher (2007) *The American Counterculture*. Edinburgh: Edinburgh University Press.

Gelder, Ken (2005) 'The Field of Subcultural Studies'. In Gelder, Ken (ed.) *The Subcultures Reader*. Abingdon: Routledge. Pp. 1–15.

Gilbert, David (2006) '"The Youngest Legend in History": Cultures of Consumption and the Mythologies of Swinging London'. *The London Journal*, 31(1): 1–14.

Goudie, A. (2013) *The Human Impact on the Natural Environment: Past, Present and Future*. Chichester: Wiley-Blackwell.

Grindrod, J. (2013) *Concretopia: A Journey Around the Rebuilding of Postwar Britain*. Brecon: Old Street.

Gruffudd, Pyrs (1994) 'Back to the Land: Historiography, Rurality and the Nation in Interwar Wales'. *Transactions of the Institute of British Geographers*, 19(1): 61–77.

Halfacree, Keith (2006) 'From Dropping Out to Leading On? British Counter-Cultural Back-to-the-Land in a Changing Rurality'. *Progress in Human Geography*, 30(3): 309–336.

Hall, Stuart (1968) *The Hippies: An American 'Moment'*. Birmingham: Centre for Contemporary Cultural Studies.

Hall, Stuart and Jefferson, Tony (eds.) (2006) *Resistance Through Rituals: Youth Subcultures in Postwar Britain*. Abingdon: Routledge.

Hebdige, Dick (1981 [1979]) *Subculture: The Meaning of Style*. London: Routledge.

Hewitt, Guy (2020) 'The Windrush Scandal: An Insider's Reflection'. *Caribbean Quarterly*, 66(1): 108–128.

Highmore, Ben (ed.) (2002) *The Everyday Life Reader*. London: Routledge.

Jackson, Cecile (2012) 'Speech, Gender and Power: Beyond Testimony'. *Development and Change*, 43(5): 999–1023.

Jones, Bryn and O'Donnell, Mike (eds.) (2010) *Sixties Radicalism and Social Movement Activism: Retreat or Resurgence?*. London; New York: Anthem Press.

Jones, Rupert (2020) 'Homebuyers "Plotting Move to Country" amid Increased Home Working'. *The Guardian*. Available at: www.theguardian.com/money/2020/may/08/homebuyers-plotting-move-tocountry-amid-increased-home-working. Accessed 21 October 2020.

Joseph-Salisbury, Remi (2020) '"The UK Is Not Innocent": Black Lives Matter, Policing and Abolition in the UK'. *Equality, Diversity and Inclusion*.

Kersen, Thomas Michael (2021) *Where Misfits Fit: Counterculture and Influence in the Ozarks*. Jackson, MS: University Press of Mississippi.

Kirk, Andrew G. (2007) *Counterculture Green: The Whole Earth Catalog and American Environmentalism*. Lawrence, KS: University Press of Kansas.

Knights, Sam (2019) 'Introduction: The Story So Far'. In Farrell, Clare, Green, Alison, Knights, Sam and Skeeping, William (eds.) *This Is Not a Drill: An Extinction Rebellion Handbook*. London: Penguin. Pp. 9–13.

Land, Victoria and Kitzinger, Celia (2011) 'Categories in Talk-in-interaction: Gendering Speaker and Recipient'. In Speer, Susan A. and Stokoe, Elizabeth (eds.) *Conversation and Gender*. Cambridge: Cambridge University Press. Pp. 48–63.

Lutfi, Nafisatul (2015) 'The Hippies Identity in the 1960s and Its Aftermath'. *Rubikon*, 2(1): 42–53.

MacLeod, Gordon (2018) 'The Grenfell Tower Atrocity: Exposing Urban Worlds of Inequality, Injustice, and an Impaired Democracy'. *City*, 22(4): 460–489.

Marsh, Jan (1982) *Back to the Land: The Pastoral Impulse in England, From 1880–1914*. London: Quartet.

Martinez, Miguel (2007) 'The Squatters' Movement: Urban Counter-Culture and Alter-Globalization Dynamics'. *South European Society & Politics*, 12(3): 379–398.

Matless, David (2016) *Landscape and Englishness*. London: Reaktion Books.

McCleary, J. B. (2004) *The Hippie Dictionary: A Cultural Encyclopedia of the 1960s and 1970s*. Berkeley: Ten Speed Press.

McKay, George (1996) *Senseless Acts of Beauty: Cultures of Resistance since the Sixties*. London: Verso.

McKay, George (2005) 'The Social and (Counter-) Cultural 1960s in the USA, Transatlantically'. In Grunenberg, Christoph and Harris, Jonathan (eds.) *Summer of Love: Psychedelic*

Art, Social Crisis and Counterculture in the 1960s. Liverpool: Liverpool University Press & Tate Liverpool.
McLean, Kate C., Pasupathi, Monisha and Pals, Jennifer L. (2007) 'Selves Creating Stories Creating Selves: A Process Model of Self-Development'. *PSPR*, 11(3): 262–278.
McRobbie, Angela (2000) *Feminism and Youth Culture* (2nd ed.). London: Macmillan.
Miller, Timothy (1999) *The 60s Communes: Hippies and Beyond*. New York: Syracuse University Press.
Miller, Timothy (2011 [1991]) *The Hippies and American Values*. Knoxville, TN: The University of Tennessee Press.
Nelson, Elizabeth (1989) *The British Counter-Culture, 1966–73: A Study of the Underground Press*. Basingstoke: Macmillan.
Newman, Loveday, Stoner, Charlotte and Spector, Aimee (2019) 'Social Networking Sites and the Experience of Older Adult Users: A Systematic Review'. *Ageing & Society*, 1–26.
Newton, Janice (1988) 'Aborigines, Tribes and the Counterculture'. *Social Analysis*, 23: 53–71.
Nicholson, Virginia (2019) *How Was It For You? Women, Sex, Love and Power in the 1960s*. London: Viking.
Oxford English Dictionary (2020a) 'Counter-Culture'. Available at: www.oed.com/view/Entry/42745?redirectedFrom=counterculture#eid. Accessed 21 October 2020.
Partridge, Christopher (2018) 'A Beautiful Politics: Theodore Roszak's Romantic Radicalism and the Counterculture'. *Journal for the Study of Radicalism*, 12(2): 1–34.
Pearson, Cynthia and Delatte, Norbert (2005) 'Ronan Point Apartment Tower Collapse and Its Effect on Building Codes'. *Journal of Performance of Constructed Facilities*, 19(2): 172–177.
Pepper, D. (1991) *Communes and the Green Vision: Counterculture, Lifestyle and the New Age*. London: Green Print.
Pile, Steve and Keith, Michael (eds.) (1997) *Geographies of Resistance*. London: Routledge.
Pred, Allan (1981) 'Social Reproduction and the Time-Geography of Everyday Life'. *Geografiska Annaler Series B, Human Geography*, 63(1): 5–22.
Rigby, A. (1974a) *Alternative Realities: A Study of Communes and Their Members, International Library of Sociology*. London: Routledge and Kegan Paul.
Rigby, A. (1974b) *Communes in Britain*. London: Routledge.
Risch, William Jay (2005) 'Soviet "Flower Children". Hippies and the Youth Counter-Culture in 1970s L'viv'. *Journal of Contemporary History*, 40(3): 565–584.
Roszak, Theodore (1969) *The Making of a Counter Culture: Reflections on the Technocratic Society and Its Youthful Opposition*. Berkley; Los Angeles: University of California Press.
Rycroft, Simon (2002) 'The Geographies of Swinging London'. *Journal of Historical Geography*, 28(4): 566–588.
Rycroft, Simon (2003). 'Mapping Underground London: The Cultural Politics of Nature, Technology and Humanity'. *Cultural Geographies*, 10(1): 84–111.
Rycroft, Simon (2007). 'Towards an Historical Geography of Nonrepresentation: Making the Countercultural Subject in the 1960s'. *Social & Cultural Geography*, 8(4): 615–633.
Rycroft, Simon (2011) *Swinging City: A Cultural Geography of London, 1950–1974*. Farnham; Burlington: Ashgate Publishing.
Samuel, Raphael (2012) *Theatres of Memory: Past and Present in Contemporary Culture*. London; New York: Verso.
Schudson, Michael (1995) 'Dynamics of Distortion in Collective Memory'. In Schacter, Daniel L. (ed.) *Memory Distortion: How Minds, Brains, and Societies Reconstruct the Past*. London; Cambridge: Harvard University Press.
Sharp, Joanne P., Routledge, Paul, Philo, Chris and Paddison, Ronan (eds.) (2000) *Entanglements of Power: Geographies of Domination/Resistance*. London: Routledge.

Sheail, J. (2002) *An Environmental History of Twentieth-Century Britain*. Basingstoke: Palgrave.

Shields, Rob (1991) *Places on the Margin: Alternative Geographies of Modernity*. London: Routledge.

Thomson, Alistair (2011) 'Life Stories and Historical Analysis'. In Gunn, Simon and Faire, Lucy (eds.) *Research Methods for History*. Edinburgh: Edinburgh University Press. Pp. 101–117.

Thrift, Nigel J. (1981) '"Owners' Time and Own Time" the Making of a Capitalist Time-Consciousness. 1300–1880'. In Pred, Allan (ed.) *Space and Time in Geography*. Stockholm: C W K Gleerup. Pp. 56–84.

Thrift, Nigel J. (1983) 'On the Determination of Social Action in Space and Time'. *Environment and Planning D: Society and Space*, 1: 23–57.

Thrift, Nigel J. (1984) *Spontaneity or Order? Time Consciousness in Medieval and Early Modern England*. Cambridge: Cambridge University Press.

Warne, Chris (2007) 'Bringing Counterculture to France: *Actuel* Magazine and the Legacy of May '68'. *Modern & Contemporary France*, 15(3): 309–324.

Weart, Spencer R. (2012) *The Rise of Nuclear Fear*. Cambridge, MA; London: Harvard University Press.

Whiteley, G. (2010) '"New Age" Radicalism and the Social Imagination: Welfare State International in the Seventies'. In Forster, L. and Harper, S. (eds.) *British Culture and Society in the 1970s: The Lost Decade*. Newcastle-upon-Tyne: Cambridge Scholars Publishing.

Willis, Paul (1977) *Learning to Labour: How Working Class Kids Get Working Class Jobs*. Farnborough: Saxon House.

Wuebbles, D. J. (2012) 'Celebrating the "Blue Marble"'. *EOS Transactions American Geophysical Union*, 93(49).

Yinger, J. Milton (1960) 'Contraculture and Subculture'. *American Sociological Review*, 25(5): 625–635.

Yinger, J. Milton (1982) *Countercultures: The Promise and Peril of a World Turned Upside Down*. London: Collier Macmillan Publishers.

Yoffe, Mark (2020) 'Soviet Rock Collection and International Counterculture Archive at the Global Resources Center of the George Washington University Libraries'. *Slavic & East European Information Resources*, 21(1–2): 112–132.

Part 1

Motivations

The chapters contained in Part 1 lead us through the start of the countercultural migrants' story. They cover three topics: why these migrants left their original homes, why they chose to move to Wales specifically, and the process of finding a home once they had made the decision to move. The stories told here are broad-ranging, introducing a wide selection of migrant experiences. Not only that, but they also introduce us to the major historical trends and events of the era. These cover a surprisingly diverse set of topics, from elections, strikes, and the IRA through to the rise of do-it-yourself (DIY) and the unexpected influence of the Ideal Homes Exhibition.

These stories are all presented together, and the chapters tell a clear chronological story. But it is important to remember that the exact timeframes vary by individual. They did not all move in the same year, nor did they all take the same amount of time to go through each stage of the process. Some migrants took years; for others, it happened very quickly. Likewise, some migrants moved right at the start of the counterculture movement, while others caught only the tail end. This means that, although presented chronologically, the historical trends and events discussed did not happen at the same point in every narrative.

At its heart, countercultural migration into rural Wales was a form of counterurban migration. Countercultural counterurban migration is not something that has received much academic attention so far. In one of the few articles on the topic, Keith Halfacree argues that 'these gaps need filling in order to appreciate the full flowering of the phenomenon in specific socio-cultural, geographical and historical contexts; to enable their relationships with contemporaneous rural socio-cultural restructuring' (2009: 771). Despite this absence of attention to the counterculture, research into counterurbanism has drawn on a wide range of different perspectives and interpretations, reflecting the complex identities commonly possessed by counterurbanisers (Phillips 2010: 539; Halfacree 2012: 210; see also, Herslund 2011; Gkartzios 2013; Adamiak et al. 2017). Some research draws on similar themes to those discussed in the following chapters, including Mitchell (2004), Walford and Stockdale (2015), and Stockdale (2016). Some original studies from the 1970s are also close, though they lack the benefit of including personal accounts (for example, Hart 1970; Jones 1965). In the chapters which follow, I build on the same theoretical ideas as this existing work, drawing together concepts around ruralism

DOI: 10.4324/9781003358671-2

and the agrarian ideal with the complexities of shifting relationships with capitalism and the desire for lifestyle change.

Counterurbanism is widely regarded as a form of lifestyle migration. In many ways, the experiences of the migrants described in the following chapters are most closely aligned with lifestyle migration. While counterurban research has gone some way towards looking at the experiences of migrants themselves, lifestyle migration takes this further. This is a relatively new concept, originating in the 1990s, and so there has been some debate over how best to define it (Swaffield and Fairweather 1998; Walmsley et al. 1998; Croucher 2015: 162). Nigel Walford and Aileen Stockdale note that 'the underlying concepts embedded in this term means that it incorporates or even subsumes earlier migration research' (2015: 103). Michaela Benson and Karen O'Reilly reached a similar concussion, suggesting that:

> lifestyle migration is not intended to demarcate and define a particular group of migrants, but rather to provide an analytical framework for understanding some forms of migration and how these feature within identity-making, and moral considerations over how to live.
>
> (2009: 21)

There is a division to be made between the ideas behind lifestyle migration research and those behind counterurbanism research. While the latter primarily grew out of an interest in *where* people were going, lifestyle migration focuses on: *why* they were moving, what lifestyle were they seeking, and what was its relationship to the place they chose to move to? Likewise, lifestyle migration research seeks to understand how successfully lifestyle migrants manage to realise their ambitions for the move. In her ethnographic account of British lifestyle migration to rural France, Michaela Benson looks not just at the thought processes that contributed to the move but also at how the migrants continued to search for better, more authentic ways of living afterwards (2011).

Since the majority of lifestyle migrants came to Wales from other parts of the UK, I draw not only on lifestyle and counterurban migration research but also on the approach taken by internal migration research. This tends to fall into two broad categories: historic and contemporary. Research into historical internal migration movements has often focused on the urbanisation of Britain. Peter Scott, for instance, explored the impact of long-distance internal migration on expanding industrial centres in interwar Britain, focusing on their social and economic impacts, quite the opposite of the issues I discuss in the following chapters (Scott 2000; see also, Pooley and Turnbull 1998; Baines 1985). Internal migration research remains very much centred on the longstanding traditions of research into population demographics, relying on statistical analysis rather than personal stories. It tends to pay little attention to migrant identities, rarely discussing internal migration in relation to topics like identity, class, gender, or equality (Smith et al. 2016: 165).

The divisions between these different forms of migration research are somewhat arbitrary. People can fit into multiple migration categories simultaneously, and this is beginning to be recognised in migration research. With this in mind, it is possible to discern how the influence of lifestyle migration as an overarching concept has impacted recent research into internal migration. In the following chapters, I build on all three forms of migration research (counterurban, lifestyle, and internal), focusing on the personal experiences of migrants as well as placing them in their wider historical (and, by extension, political and cultural) context.

References

Adamiak, C., Pitkänen, K. and Lehtonen, O. (2017) 'Seasonal Residence and Counterurbanisation: The Role of Second Homes in Population Redistribution in Finland'. *Geojournal*, 82: 1035–1050.

Baines, Dudley (1985) *Migration in a Mature Economy: Emigration and Internal Migration in England and Wales, 1861–1900*. Cambridge: Cambridge University Press.

Benson, Michaela (2011) *The British in Rural France: Lifestyle Migration and the Ongoing Quest for a Better Way of Life*. Manchester: Manchester University Press.

Benson, Michaela and O'Reilly, Karen (2009) 'Migration and the Search for a Better Way of Life: A Critical Exploration of Lifestyle Migration'. *The Sociological Review*, 57(4): 608–625.

Croucher, Sheila (2015) 'The Future of Lifestyle Migration: Challenges and Opportunities'. *Journal of Latin American Geography*, 14(1): 161–172.

Gkartzios, Menelaos (2013) '"Leaving Athens": Narratives of Counterurbanisation in Times of Crisis'. *Journal of Rural Studies*, 32: 158–167.

Halfacree, Keith (2009) '"Glow Worms Show the Path We Have to Tread": The Counterurbanisation of Vashti Bunyan'. *Social & Cultural Geography*, 10: 771–789.

Halfacree, Keith (2012) 'Heterolocal Identities? Counter-Urbanisation, Second Homes, and Rural Consumption in the Era of Mobilities'. *Population, Space and Place*, 18: 209–224.

Hart, R. A. (1970) 'A Model of Inter-Regional Migration in England and Wales'. *Regional Studies*, 4(3): 279–296.

Herslund, Lise (2011) 'The Rural Creative Class: Counterurbanisation and Entrepreneurship in the Danish Countryside'. *Sociologia Ruralis*, 52(2): 235–255.

Jones, H. R. (1965) 'A Study of Rural Migration in Central Wales'. *Transactions of the Institute of British Geographers*, 37: 31–45.

Mitchell, Clare (2004) 'Making Sense of Counterurbanisation'. *Journal of Rural Studies*, 20: 15–34.

Phillips, M. (2010) 'Counterurbanisation and Rural Gentrification: An Exploration of the Terms'. *Population, Space and Place*, 16: 539–558.

Pooley, Colin and Turnbull, Jean (1998) *Migration and Mobility in Britain since the Eighteenth Century*. London: UCL Press.

Scott, Peter (2000) 'The State, Internal Migration, and the Growth of New Industrial Communities in Inter-War Britain'. *The English Historical Review*, 115(461): 329–353.

Smith, Darren P., Finney, Nissa, Halfacree, Keith and Walford, Nigel (2016) 'Conclusion: Moving Forward'. In Smith, Darren P., Finney, Nissa and Walford, Nigel (eds.) *Internal Migration: Geographical Perspectives and Processes*. London: Routledge. Pp. 165–178.

Stockdale, Aileen (2016) 'Contemporary and "Messy" Rural In-Migration Processes: Comparing Counterurban and Lateral Rural Migration'. *Population, Space and Place*, 22: 599–616.

Swaffield, S. and Fairweather, J. (1998) 'In Search of Arcadia: The Persistence of the Rural Idyll in New Zealand Rural Subdivisions'. *Journal of Environmental Planning and Management*, 41(1): 111–118.

Walford, Nigel and Stockdale, Aileen (2015) 'Lifestyle and Internal Migration'. In Smith, Darren P., Finney, Nissa, Halfacree, Keith and Walford, Nigel (eds.) *Internal Migration: Geographical Perspectives and Processes*. London: Routledge. Pp. 99–112.

Walmsley, D. J., Epps, W. R. and Duncan, C. J. (1998) 'Migration to the New South Wales North Coast 1986–1991: Lifestyle Motivated Counterurbanisation'. *Geoforum*, 29(1): 105–118.

1 Why leave? Reasons for joining the countercultural migration

In this chapter, we meet the six central figures in this research and begin to explore the stories they shared. This is where we start to see how, by analysing their stories, we can gain a sense of their experiences. This is not a new approach to research but a recognised function of narrative identity. As Jefferson Singer has observed, 'to understand the identity formation process is to understand how individuals craft narratives from experiences, tell the stories internally and to others, and ultimately apply these stories to knowledge of self, other and the world in general' (2004: 438).

Dan P. McAdams developed his theory of narrative identity in 1985, and it has gone on to become deeply influential across a range of disciplines. He and Kate C. McLean define narrative identity as 'a person's internalized and evolving life story, integrating the reconstructed past and future to provide life with some degree of unity and purpose' (2013: 233). Echoes of these ideas come up regularly in work on identity and place (see, for instance, Christou 2006; Leyshon and Bull 2011; Riley and Harvey 2007; Tyrrell and Harmer 2015). Sociologist Margaret Somers has also argued in favour of narrative identity theories, suggesting that they offer a way of avoiding an overemphasis on strict categories (1994). Since the early 2000s, there has been an increase in awareness of how the stories we tell about our lives embody more than just our view of ourselves, with Aman Sium and Eric Ritskes arguing that 'stories and storytelling are political, always more than personal narratives' (2013: v–vi). Keven Hetherington draws a crucial distinction between understanding categories, such as public/private or urban/rural, as fixed and understanding them 'as processes that are made through social practices rather than already existing as social forms' (2000: vii–viii).

The stories in this chapter and throughout the book are therefore more than simple recollections of past events. Instead, these oral histories provide nuanced and detailed information on people's perceptions of themselves, their lives, and the world they inhabit. This chapter focuses on what pushed these migrants to move to Wales and become part of the counterculture (however they understood the term) and puts these experiences into their historical context. The aim is to build up a multi-layered picture of the world at the start of our grand narrative and show how the individual experiences and actions of the migrants both contributed to and were created by their surroundings. As established in the Introduction, there is a

DOI: 10.4324/9781003358671-3

stereotypical image of hippy migrants and their reasons for living the life they do. I argue that this image is too narrow and does not take into account the diversity of people's experiences.

1.1 Escape from convention

During the post-war years in Cambridge, a little girl named Lesley was busy growing up (as all children do). Lesley became a vegetarian when she was still small, displaying an interest in alternative lifestyles from an early age. Her passion for self-sufficiency and fascination with traditional peasant lifestyles grew with her. While her extended family teased her, dubbing her 'Farmer Lesley', her close family encouraged her unusual interests:

> I can remember way back when I was a kid my sister and I had this train set, quite a big train set. We weren't brought up as little girlies, we had a train set. And on her side of the train set she had this like, nice little sort of village with nice houses, I can remember making these little huts, like little wattle and daub huts for my peasants to be in. And yes, I must've only been about six, seven, eight, but already that sort of thing about that sort of really simple basic lifestyle. And I used to subscribe to a magazine when I was again, it must've been pre-politics so it must've been around sort of 11, 12, called the, *The Smallholder*. I used to get sort of books out of the library on sorts of breeds of cattle and chickens and stuff like that. So it really was a very strange and strong thread.

At 13, Lesley encountered Buddhism for the first time, and she immediately converted, relying on what information she could find in books to learn meditation techniques. The pull towards a peaceful, self-sufficient, peasant lifestyle was strong from the beginning.

The other half of Lesley's life growing up was politics. Her family were active members of the Labour Party, and she joined Young Labour as soon as she was old enough. As time went on, though, she became disenchanted with the political scene. It seemed to be constantly failing to bring about any real structural change in society or to rectify the chaos that was ensuing as the perceived stability of the 1960s gave way to the outright turbulence of the 1970s. Politics was failing to provide the future that she was raised to believe it would, and so she found herself turning to her passion for self-sufficiency for answers. It was this shift in focus which led to the concoction of her grand plan. Lesley and her then-partner decided that they would travel the hippy trail to India, and then when they returned, they would move to Wales and begin living a self-sufficient lifestyle:

> It was so clear we were going to do this trip to India then we were going to move to Wales, buy somewhere and grow vegetables and stuff like that. . . . There was a clear plan. Um, maybe the clearest plan I've ever had in my life actually.

In many ways, Lesley's experiences were those of the archetypal hippy incomer. Being always the odd one out, never quite fitting into conventional society, and showing hints of rebellion in a dissatisfaction with mainstream politics – these are the hallmarks of the hippy stereotype. Lesley chose to adopt this lifestyle because she saw living by example as the best way of promoting change, where politics had so far failed to do so. But she also wanted to do it because it was something she'd had a passion for since childhood. Regardless of the potential benefits to humanity, it was a chance to live out her dreams.

Lesley had no interest in living a consumerist lifestyle and adopted a 'peasant' lifestyle as soon as she was able to, but the big push towards this was the failure of the political system to maintain order. John O'Farrell remarks that:

> the cultural and social history of the [60s] is so appealing to anyone looking back that the true story has tended to be obscured: that it was a time of almost unremitting economic crisis, of desperate short term financial fixes and cobbled together Treasury cutbacks.
>
> (2009: 133)

This chaos continued into the 1970s, as life was repeatedly upended by a succession of strikes, the oil crises, IRA violence, and the constant underlying threat of nuclear war (Grindrod 2013: 363). The Cuban Missile Crisis in October 1962 gripped and terrified the world and kicked off the era of fear around nuclear threats which would continue for decades (DeGroot 2009: 78). Likewise, it was in the 1970s that the environmental threat posed by increasing economic growth was first recognised by scientists with concern (Wall 2005: 70). With so many disasters looming, the dominant political parties seemed to be failing to address any of them, and British politics itself was on shaky ground. In 1974, the Conservatives won their lowest share of the vote since 1918, but in the five years which followed, their fortunes saw a remarkable reversal, a testament to the rapid changes to which the decade was witness (Saunders 2012: 27). The result of this tumult was that people like Lesley felt the system wasn't working for them. Roger King wrote at the close of the 1970s that members of the middle class, like the family that Lesley grew up in, had spent the decade feeling ignored by politicians. In the ten years between 1964 and 1974, trade union membership among middle-class professions increased from 29% to 40% (King 1979: 2 & 7).

It was in the search for another alternative to mainstream politics that people like Lesley joined, and in so doing, helped to create the alternative movement. Roger King went on to describe how

> the 'middle class revolters' robustly recognise their arch-enemies, and are rarely reluctant to condemn them publicly. . . . Groups clearly associated with the detested state machinery, such as civil servants, politicians, planners, and their cultural allies in education and the mass media, are persistent objects of anger.
>
> (1979: 18)

This 'detested state machinery' is what seemed to be failing and what the alternative movement was trying to reinvent. It is something we see most clearly in the development of communes, but independent movers like Lesley are part of the same trend.

Around the same time Lesley was busy growing up in Cambridge, a girl named Barbara was also busy growing up in a small rural community in the west of Northern Ireland. Like Lesley, she became interested in sustainable ways of living, adopting a vegetarian diet while still young. But unlike Lesley, alongside this developing interest in alternative lifestyles, she became increasingly irritated by the place where she lived. Barbara found life in the archetypal small rural community repressive. As she would reflect later, '[I] couldn't wait to get away from the area really, because it was very stifling and everybody knew everybody else'. Like generations of young people from the countryside before her, Barbara was initially on the lookout for a way to escape rural life. Eventually, she fled to Dublin, where she studied for a BA in Modern Languages at university. Here, away from the limited potential of her hometown, she was free to explore her interests and mix with a far wider variety of people. Through befriending students from America, she was introduced to American advancements in macrobiotics and self-sufficiency, continuing the interest in alternative living that she developed as a child. Following university, she worked as a lecturer in France, and after a few years there, she moved to London.

It was in London that things started to become difficult. The consumerist nature of life in the capital made it hard for her to settle there; it clashed with her belief in the importance of sustainability. Despite the fact that at this time approximately one in eight British people lived in London, the self-centredness of consumerist lifestyles made it an isolating place where nobody seemed to communicate properly in the way they had in the country (Benson 1994: 16). This was not just something Barbara noticed or, indeed, something that was unique to London. In a 1978 interview, a resident of Caldmore, a suburb of Walsall in the Black Country, commented that 'the biggest change I've seen here over the years has been an increase in competitiveness, people's rivalry with each other. People don't know each other as they used to' (Seabrook 1978: 79–80). Despite having hated rural life when she was a child, Barbara now finds her thoughts returning to it:

> Came back to London and was very struck when I got back, got to live in London at how consumer orientated it was, how busy it was, how everybody was out for the main chance. And again although I'd fled from rural Ireland as it were, I was very struck by the fact that people didn't communicate at all really. So it was, although I'd found it stifling as a child it was, you know, part of my make-up really. And I found it really difficult to be in London, I hated it. So the mantra almost came to be 'we need to get away from this, we've got to get away from it'. And came to Wales in 1970 to visit a college friend who I'd shared a house with in university, and actually came to Aberystwyth to visit them. And was really struck by the peace, the quiet, and the emptiness really. The village that we came to visit, um, which is where I live now, there were a lot of empty houses. A lot of, it seemed to me that

a lot of the young people were leaving Wales, maybe for the same reasons that I'd fled Ireland, I don't know. Um, but there was a lot of, there were a lot of empty houses. And they, in the village that I eventually moved to there were really empty houses and no people really. So, and also there was land that was available with these houses, so it seemed to be a really nice idea to just try and get out of London, move somewhere rural, and maybe pursue the whole get away from it, self-sufficiency, um, respect for the planet, that kind of thing.

In the 1960s, the end of the Age of Austerity had led to a boom in consumerism, Britain's manufacturing industry was prospering, and Prime Minister Harold Macmillan proudly proclaimed that Britain had 'never had it so good' (Sandbrook 2005). For those who were happy to participate in this new world, that may well have been true, but those who were wary of the sudden increase in consumption and production had to find an alternative, and fast. Humans have always been consumers, but we have not always been plagued by overconsumption. John Naish wrote that 'in the 1950s happiness was still being sold as a helpful personal habit, but by the 1970s aspiration inflation had set in and we were being told that we were entitled to be happy' (2008: 172). With rising wages and relatively low inflation levels, the average expendable income grew dramatically between the 1950s and 1970s, and with more money to spend on non-essentials, people were far more likely to be drawn into a consumerist lifestyle (Saville and Gardiner 1986: 52; Benson 1994: 13). Herbert Marcuse, an influential figure in developing the philosophy of the counterculture movement, wrote in 1964 that:

the people recognize themselves in their commodities; they find their soul in their automobile, hi-fi set, split level home, kitchen equipment. The very mechanism which ties the individual to his society has changed, and social control is anchored in the new needs which it has procured.

(9)

More so than ever before, people were constructing and expressing their identities and relationships through their belongings. Not just that, but they were relying increasingly on big corporations to do so. Shopping habits were changing rapidly, with large American-inspired self-service supermarkets proliferating for the first time (Shaw et al. 2004: 574). In 1939, the average Briton did 60% of their shopping at small independent stores. By 1981, this figure had dropped to just 31% (Benson 1994: 62). This was particularly significant for women; as shopping is a historically female domain, the task of buying food and other necessities has overwhelmingly fallen to the women of the household (Bowlby 2000: 7).

When consumerism surrounded you, it wasn't always possible to avoid it, even if you wanted to. Historian Virginia Nicholson recalls how:

Though my parents (like many of their high-minded generation) were resistant to watching commercial television, its consumerist agenda filtered

> through, subtly reshaping their lives. The adverts for After Eight Mints and Cadbury's Drinking Chocolate, for Vesta Chow Mein and Heinz Sandwich Spread all worked their magic on us and we wanted our share.
>
> (2019: 265)

For anyone like Barbara who was disinterested in becoming part of this consumer culture, this resulted in a sense of isolation and a lack of connection: other people were interacting in a way that she (and others) could not relate to. This is a recognised problem which occurs when people cannot or will not engage with consumerism on the same level as the social groups they interact with, and because women are more likely to shop, they are more likely to succumb to this, just as Barbara did (Manktelow 2011: 260). Living in an urban environment can also lead to a lack of social connection, which is again evidenced in Barbara's story. Stressful environments, such as those which are noisy, crowded, or polluted (all characteristics of urban life), can lead to social isolation (Wandersman and Nation 1998: 651). This is what pushed Barbara to leave London: the lack of connection and social isolation that living a consumerist lifestyle in an urban environment leads to. This gives a different perspective on the archetypal story of anarchistic hippies drawn to self-sufficiency. As Barbara's story shows, not everyone was automatically comfortable with the idea of rural life, and not everyone was coming from the same point in their lives; the migrants varied by age, life stage, nationality, and so on.

Josie had already made the decision to leave London when her thoughts turned to Wales, and was living in Afghanistan while running an export business with her husband at the time. As a child growing up in St Albans, Josie was plagued by a constant feeling of never quite fitting in, disinterested as she was in the burgeoning consumer culture. She moved from St Albans to London at 17 after leaving school, but the move did not quite pan out:

> I always felt like a square peg in a round hole. I never felt like I fitted, I didn't, I wasn't interested in doing the consumer thing, I was wanting a different way of life. As many were my age, you know, I was born in 1947, so part of that post-war baby boom thing. And we saw things very differently, we were the first teenagers to have, um, money, and there was a whole music thing, and you know, so going back a little bit and in 1967 I was part of the Flower Power movement, I was in London for all of that.

London was the cultural capital of the world at this time. As a film critic and historian, Shawn Levy describes:

> The 50s were Paris and Rome.
>
> The '70s California, Miami and New York.
>
> But the '60s, that was Swinging London – the place where our modern world began.
>
> (2002: 6)

It was partly the Second World War which led to London's rise in fashionability. It had emerged victorious, never having fallen to fascism or communism, and the destruction it suffered during the Blitz had paved the way for modern redevelopment by freeing up space throughout the city (Mort 2010: 7). A great deal of the youth scene in London was driven by consumerism, the same phenomenon that Josie (like Barbara) was disinterested in. Between 1938 and 1958, the wages of adolescents rose twice as quickly as those of adults, so by the time Josie arrived in London, it was full of young adults with prodigious spending power (Benson 1994: 18). It helped, too, that the overall number of young people had increased as a result of the post-war baby boom. Simon Rycroft describes how much of London's reinvigorated image and culture came about because of this, with a 20% increase in the numbers of 15–19-year-olds between 1956 and 1963 (2011: 68).

It was while in London that Josie became interested in macrobiotics, just as Barbara had done in Dublin. The macrobiotic diet was developed in Japan during the nineteenth century, and it prioritised eating unrefined foods, whole grains, and fresh seasonal produce (The Macrobiotic Association). It was introduced to Britain in the 1960s by the American brothers Craig and Greg Sams, who opened the first macrobiotic shop and café in London (ibid). Josie met the Sams brothers and was a frequent customer, adopting a vegetarian diet as a result. Vegetarianism was not a new phenomenon, but vegetarianism in the 1960s and 1970s differed from that which had come before it. When it was founded in the mid-nineteenth century, the Vegetarian Society was opposed to not just meat but also fermented products and anything that they considered a stimulant. This ruled out most spices and herbs, as well as tea and coffee. Countercultural vegetarians of the 1960s, on the other hand, were concerned with eating peacefully ethically produced food that was organic and unrefined (Spencer 1993: 333–334).

In ways like this, Josie and her husband were immersed in the counterculture, but these peaceful, ethical values did not fit well with the reality of city life. They decided that they would go travelling in a bid to escape the constraints of society, a common activity for countercultural enthusiasts at the time. Travel was the easiest form of escape and offered a quick, clean break with the conventions of London's consumer society. The 1960s were the first time that significant numbers of British teenagers and young people began travelling abroad, and it was partly the growth in youth income which enabled these travels to happen (Benson 1994: 95). This meant that the hippy trail originated, to some extent, from the same roots as the consumer culture it was created to escape from. The precise route fluctuated but followed the same basic pattern: 'India was usually the ultimate destination, with stop-overs along the way in Greece, Turkey, Iran and Iraq, in Afghanistan, Nepal and Pakistan. Only a small minority made it even further, to Burma, Thailand and, for gore-trippers, Vietnam' (Green 1998: 226). Because this was still the early days of international travel and few guidebooks on the hippy trail existed, people relied on information transmitted by word-of-mouth to decide where to go and what to do (ibid: 225). Josie and her husband ended up in Kabul, where they took care of a friend's business and eventually went on to start their own handcraft export company.

Things changed once their first daughter was born while they were still living in Afghanistan. Josie became pregnant with their second soon afterwards, and with children to think about, they found their feelings about travelling changing:

> I wanted to put down roots, I didn't want to just be travelling anymore I wanted to put down roots, so we came home, and we'd saved up some money, and we looked all around for somewhere to buy as places were expensive, and we found somewhere in Lampeter. And it was interesting because while I was in Afghanistan my friend was living in this area and she had been writing me letters, and the letters seemed to just drip greenery, and Afghanistan was having an eight-year drought. So you know, I thought, I felt like I already had a connection with this area?

Josie wanted to settle down with her family, but she was also looking for somewhere that they could continue to live a countercultural lifestyle. She did not want to be re-immersed in consumerism, and she wanted to be able to continue her interest in macrobiotics and vegetarianism. Unlike Lesley and Barbara, she had already made the break with conventional society when she arrived in Wales. It was the arrival of her daughters that pushed her and her husband towards a settled lifestyle.

Josie's story illustrates two separate tipping points that led to her arrival in Wales. The first was the decision to leave London. Josie left the city when the chaos of the 1970s had not yet hit properly. It was the growing consumerism that she was getting away from – the way of life that became the standard for city living – the same culture that Barbara was finding she could not deal with either and which Lesley had never even tried to adopt. The second tipping point was the decision to leave Afghanistan. This was fundamentally different from the decision to leave London because the commitment to a countercultural lifestyle had already been made. If anything, it is closer to a step back towards normality and certainly could have become so if Josie and her husband had not been so determined not to let it. There are elements of her dislike of city life creeping back to the surface of her narrative here, despite the fact that Kabul is a very different kind of city from London. She recalled how 'in Afghanistan the cows would graze on rubbish . . .', an image reminiscent of the rubbish that was simultaneously building up in London's strike-ridden streets.

Just from these three examples, we have already begun to see how the stories of people who moved are fantastically diverse, nuanced, and interconnected. While on the surface it may seem that people were pushed out of cities for similar reasons – Lesley, Barbara, and Josie's stories all contain a dislike of consumerism, for example, and a reaction to the turbulent politics of the time – beneath this there are a plethora of personal influences as well: Josie's children, Barbara's rural upbringing, Lesley's Buddhist beliefs. These are things we do not think of in the stereotype of the anarchist hippy, someone who only exists in response to external factors. And yet Lesley, Barbara, and Josie are still quite close to this archetypal image in the grand scheme of things. They all express 'a moral rejection of the established society', as described by John McCleary in the introduction, and they are not afraid to make radical changes to put these beliefs into action (2004: 246). The next two stories will show how the stereotypes can be pulled apart even further.

1.2 Safety and freedom

Judith, a born and bred Londoner, was severely affected by the dangers that the increasing IRA violence and strikes of the 1970s had led to. From her perspective, the city she called home had become dangerous and unfriendly. Rubbish was accumulating in the streets. Bombs were going off. Her family was not safe.

> I had a son who was about 13, who was travelling across London, from north-west London to Fulham, and I was concerned that I might not see him when we all got back from work. And I suppose he might have, had he thought about it, he might have felt the same about his parents. We were part of a group of friends who seemed to be all heading off in different directions. As I mentioned earlier, one family was going to North Africa, lots of, several other friends were going to Wales . . . so there's a sense of people leaving London, a sense of London not being a very nice place to live. As I mentioned earlier, the rubbish wasn't being collected, the dead weren't being buried, and then what we called the three-day week.

Before now, life in London had been relatively happy: Judith enjoyed plenty of freedom as a child, and her daughter Catherine was benefitting from London's vibrant multiculturalism, attending a city school with children from around the world. Judith's parents were always close at hand for support. But the turbulence of the time meant that London was fast becoming an unfriendly place to live. As people she knew began packing up and leaving, she began to wonder if her family, too, would be better off elsewhere. At the same time, her husband was dreaming up plans for communal living and other ways of improving humankind's relationship with the environment. The idea of somewhere safe, where her husband could dream and plan and the children could play and be free, beckoned strongly.

Judith wanted safety for herself and her family. She wanted them to have a secure home where they could do what they wanted without having to deal with fear of their surroundings. Before now, there had not been a prompt to leave London; it had seemed a perfectly safe place to live and bring up children. It was only when things started to deteriorate in the city that Judith felt the need to leave. This was also what pushed her husband to take the plunge and focus on his ideas around communal living and the environment; it was not until the city became a bad place to live that this happened. They could not move to Wales immediately and first spent time living with in-laws in Cornwall, but when a house became available in Barmouth, they jumped at the chance:

> a big house for sale became vacant, that we had been offered the opportunity to live in rent free for a year while my husband was working on his ideas

> for co-operative living. So that's the reason we came to Wales, so he could pursue what he wanted to do.

Judith was not alone in perceiving the city as more dangerous and unwelcoming than the countryside, particularly where bringing up children is concerned (Valentine 1997: 140). A plethora of negative images surround the idea of cities; they are 'a site of anomie, alienation, corruption, ill health, immorality, chaos, pollution, congestion, and a threat to social order' (Bridge and Watson 2003: 15–16). This is the flip side of the long-standing stereotype of 'rural idylls'. In the nineteenth century, Tönnies identified two types of social formation: *gesellschaft* (complex urban societies) and *gemeinschaft* (simple rural communities) (2012 [1887]: 16). These were very polite descriptors: over time, cities have been continually viewed and represented as not just complex but downright unpleasant, in contrast to the idyllic simplicity of the country.

As we have seen from Lesley, Barbara, and Josie's experiences, strikes and fuel shortages meant that in the 1970s, the chaotic, negative representations of the city were becoming a reality. As Judith recalled earlier, rubbish was being left uncollected in the street, the dead were being left unburied, and the very fabric of urban life was disintegrating. Speaking about the 1972 Miners' Strike, Arthur Scargill stated that:

> You see, we took the view that we were in a class war. We were not playing cricket on the village green like they did in '26. We were out to defeat Heath and Heath's policies because we were fighting a government. Anyone who thinks otherwise is living in cloud-cuckoo land. We had to declare war on them, and the only way you could declare war was to attack the vulnerable points. They were the points of *energy*; the power stations, the coke depots, the coal depots, the points of supply. And this is what we did.
>
> (Clutterbuck 1978: 59, emphasis in original).

These strikes were drastic bids for attention, aimed at damaging the roots of society's infrastructure. And while their aims may have been worthy and their cause justified, they made London a difficult place to raise a young family. The rising consumerism despised by Barbara and Josie also meant rising waste, and this meant that when the bin collectors also went on strike, the consequences were more severe than they would have been 20 years earlier. Dominic Sandbrook records the comments of a visitor to Blackburn in the 1970s, who saw in the bins there a multitude of waste that could only have been imagined in the past:

> bottles and jars, of milk, sauce, salad cream, pickles, jam: tins of beans, fish. Fray Bentos steak pies, fruit, spaghetti rings, Spam, vegetables; hardened slices of bread fanning from their waxed paper, silver foil trays from the takeaway, egg boxes, plastic containers, dispensers of foams, creams, oils and polish; nappies, tights, old clothing.
>
> (2013: 27)

To top it off, the first IRA bombs were planted in London in 1973. The Provisional IRA had been founded in 1969, and increasing frustration at a lack of attention meant that the mid-1970s saw attacks on the Tower of London, the Old Bailey, Euston Station, and the Prime Minister's own home, as well as countless shops, restaurants, and hotels both in the capital and outside it (Beckett 2009: 118–119). These attacks would go on to cause over 100 deaths throughout Britain, easily enough to scare people (McKittrick and McVea 2002: 91). This combination of factors all coinciding at once was what it took to push Judith and her family to leave London. Like Lesley, they were swapping this chaos for a lifestyle that they thought would be more secure and sustainable, away from the direct consequences of politics' failure to maintain order. But while Lesley was consciously choosing to set an example for others, Judith was more concerned with the safety and wellbeing of herself and her family. In this sense, her story echoes Josie's more than it does Lesley's. Judith's husband might have had some grand ideas about alternative living, but these are a secondary part of her own narrative.

For Ian, the story is different again. Ian's mother died when he was young, leaving him an only child with his father, a lecturer at City University in London. Ian, left-wing, outdoorsy, and interested in art, music, and literature, felt he had nothing in common with his right-wing scientist father. All the same, he wanted to please his father and grandfather by going to university, and so he applied to study at institutions that seemed to be in nice places and which would play to his interests: Exeter, Edinburgh, and Aberystwyth. This is how he came to live in Wales, as it was Aberystwyth's offer of a place on their English and Art degree scheme that he accepted. This choice of subject separated him from his scientist father while still tying into his pre-existing interests.

> I'd read quite a lot of books, the usual things, D.H. Lawrence, Herman Hesse, er, various authors. Music, um, mainly singer songwriters, folk influence, Bob Dylan, Dan Lawrence and Leonard Cohen, all kind of swing band, the main ones. . . . We used to go to lots of free festivals in, um, Hyde Park. And Glastonbury festivals, and then, I don't know, I went to some in Lincoln I remember. . . . Although I was in the suburbs in London I was aware of what was going on.

The seeds of interest in counterculture were there from the beginning, but they did not properly grow until after Ian's arrival in Wales. The decision to go to university was born out of a desire to please his family while escaping from them, something far more in keeping with the countercultural ethos than an academic career.

> I wanted to finish as soon as I could really, really I think I was trying to please my father and my grandfather, actually. I just wanted to finish academia as quickly as I could. . . . Yeah, I really wanted to get away from London, get away from home actually.

Students are, of course, a key part of the stereotypical hippy image. The number of students in higher education increased drastically after the Second World War. Before this, there were never more than 70,000 university students in Britain, but by 1965, there were 300,000 (Bugge 2004: 187). This is, of course, partly due to the overall increase in numbers of young people that we have already heard about in Barbara's story – as the numbers of youth grew, so did the institutions dedicated to them. It's hardly surprising, therefore, that university towns seemed to have suddenly been taken over by young people. And the number of university towns was itself increasing in the period, as rising demand for university education led to the development of ten new British universities between 1961 and 1979 (Pellew and Taylor 2021: 1). From the 1960s through the 1980s, young people became steadily more and more important, going from merely being treated as inexperienced adults to a powerful economic and cultural force (O'Farrell 2009: 158).

For Ian, the decision to move to Wales was born out of a desire to end up in a pleasant place rather than to pursue an explicitly countercultural lifestyle. Ian wanted the chance to figure out exactly what he wanted in life by going somewhere that he could immerse himself in the books and music he liked and the outdoor world he loved. He did not know exactly what he wanted just yet – just that his father's world was not it. This reflects the dissatisfaction that many are feeling with the state of modern life. Writing in 1978, journalist Jeremy Seabrook observed that:

> Most of us now do not want for basic comforts: and this has been achieved, not for the most part by exercising our skills, but by forfeiting them. Many of us resent the work we do now. We grudge the use of our time, and are often indifferent to the things we make or the services we provide. We feel bored and functionless. We see work seldom as something worthwhile in itself, but as a means to something else: it is an unhappy intrusion into the real business of our lives. We measure ourselves not by what we do, but by what we can acquire. Our function is no longer a primary determinant of our identity.
> (12)

It was only when Ian arrived in Wales and settled here that he really became properly drawn to the counterculture. Ian grew up to become a self-described hippy, and the ideals of the counterculture movement became a big part of his life. The start of his story, along with the others covered here, show that the story of the hippy incomers did not always begin where we might expect it to.

During the miners' strikes of the 1970s, Lord Chancellor Hailsham said that:

> The war in Bangladesh, Cyprus, the Middle East, Black September, Black power, The Angry Brigade, the Kennedy murders, Northern Ireland, bombs in Whitehall and the Old Bailey, the Welsh Language Society, the massacre in Sudan, the mugging on the tube, gas strikes, hospital strikes, go-slows, sit-ins, The Icelandic Cod war [are all] standing on or seeking to stand on different parts of the same slippery slope.
> (Cresswell 1996: 63)

Just as the establishment saw all of the 1970s chaos as interlinked, so we have seen throughout the course of this chapter that the things which pushed Lesley, Barbara, Josie, Judith, and Ian to abandon their existing lives are also interlinked and interwoven with each other. The events of the 1960s and 1970s, such as those described by Lord Hailsham, also played a key role. As we have demonstrated, these years saw drastic changes in British society. Consumerism was taking off, recovering from the hit it had taken during the Second World War. It rapidly changed the way people constructed their identities and relationships, driving people like Barbara and Josie to search for somewhere else to belong. Post-war rebuilding efforts modernised cities as towns embarked on slum clearance programmes and replaced them with tower blocks and managed suburbs. At the same time, discontent was leading to strikes, which caused power cuts and chaos in Britain's cities, and the IRA was embarking on a series of attacks on the mainland. This was enough to drive people like Judith away from the city and into the countryside. And the perceived failure of politics to remedy the situation meant that many took the path that Lesley did, trying to find a better way of living.

In their research on lifestyle migration, Benson and O'Reilly have created three separate typologies of lifestyle migrants, each with different motivations for moving and different aspirations for their new life. These are:

- 'The residential tourist', who chooses to make permanent homes in their favourite holiday destinations. Popular destinations for this group now include the Algarve or Costa del Sol.
- 'The rural idyll seeker', attracted by the promise of 'the good life', community spirit, and reconnection with the land. This group often moves to scenic parts of France or Spain.
- 'The bourgeois bohemian', driven by a desire to engage with creative or spiritual pursuits and different cultures. They typically end up in a wide range of places, including Greece, Italy, and India.

(2009: 611–613)

As we have seen, all of these typologies appear in some way or another in the countercultural migrants' reasons for moving. The distinctions between them are far from clear-cut in practice, but the majority of migrants seem to have combined elements of 'the rural idyll seeker' and 'the bourgeois bohemian'. They moved because of rising consumerism, inhospitable urban environments, terrorist threats, and a desire to escape and belong. But at the same time, they each brought their own nuances to the reasons for moving. On the surface, for example, Lesley, Barbara, and Josie all moved to Wales for the same reason: to escape the city and live a more fulfilling life. But as I have demonstrated here, when their stories are recounted, there are numerous subtle, personal differences between them. For Lesley, this was not just a reaction against failed politics but a chance to explore her own deep-rooted interests in sustainability and peaceful living. For Barbara, it was a return to the kind of community she had grown up in and missed. Josie and Judith both wanted safe, secure homes to raise their children. And Ian just wanted some space to find out who he was.

References

Beckett, Andy (2009) *When the Lights Went Out: What Really Happened to Britain in the Seventies*. London: Faber and Faber.

Benson, John (1994) *The Rise of Consumer Society in Britain, 1880–1980*. London; New York: Longman.

Benson, Michaela and O'Reilly, Karen (2009) 'Migration and the Search for a Better Way of Life: A Critical Exploration of Lifestyle Migration'. *The Sociological Review*, 57(4): 608–625.

Bowlby, Rachel (2000) *Carried Away: The Invention of Modern Shopping*. London: Faber and Faber.

Bridge, Gary and Watson, Sophie (2003) 'City Imaginaries'. In Bridge, Gary and Watson, Sophie (eds.) *A Companion to the City*. Oxford: Blackwell Publishers.

Bugge, Christian (2004) '"Selling Youth in the Age of Affluence": Marketing to Youth in Britain since 1959'. In Black, Lawrence and Pemberton, Hugh (eds.) *An Affluent Society? Britain's Post-War 'Golden Age' Revisited*. Aldershot: Ashgate. Pp. 185–202.

Christou, A. (2006) *Narratives of Place, Culture and Identity: Second-Generation Greek-Americans Return 'Home'*. Amsterdam: Amsterdam University Press.

Clutterbuck, Richard (1978) *Britain in Agony: The Growth of Political Violence*. London: Faber and Faber.

Cresswell, T. (1996) *In Place/Out of Place: Geography, Ideology, and Transgression*. Minneapolis; London: Minnesota University Press.

DeGroot, Gerard (2009) *The Sixties Unplugged*. London: Macmillan.

Green, Jonathan (1998) *All Dressed Up: The Sixties and the Counter-Culture*. London: Jonathan Cape.

Grindrod, J. (2013) *Concretopia: A Journey Around the Rebuilding of Postwar Britain*. Brecon: Old Street.

Hetherington, Kevin (2000) *New Age Travellers: Vanloads of Uproarious Humanity*. London; New York: Cassell.

King, Roger (1979) 'The Middle Class in Revolt?'. In King, Roger and Nugent, Neill (eds.) *Respectable Rebels: Middle-Class Campaigns in Britain in the 1970s*. London: Hodder and Stoughton. Pp. 1–22.

Levy, Shawn (2002) *Ready, Steady, Go! Swinging London and the Invention of Cool*. London: Fourth Estate.

Leyshon, Michael and Bull, Jacob (2011) 'The Bricolage of the Here: Young People's Narratives of Identity in the Countryside'. *Social & Cultural Geography*, 12(2): 159–180.

The Macrobiotic Association. Available at: https://macrobiotics.org.uk. Accessed 20 May 2019.

The Macrobiotic Association. 'History of Macrobiotics'. Available at: https://macrobiotics.org.uk/history-of-macrobiotics/. Accessed 17 January 2020.

Manktelow, Roger (2011) 'Community, Consumerism and Credit: The Experience of an Urban Community in North-West Ireland'. *Community, Work & Family*, 14(3): 257–274.

Marcuse, Herbert (1964) *One-Dimensional Man: Studies in the Ideology of Advanced Industrial Society*. London: Routledge.

McAdams, Dan P. and McLean, Kate C. (2013) 'Narrative Identity'. *Current Directions in Psychological Science*, 22(3): 233–238.

McCleary, J. B. (2004) *The Hippie Dictionary: A Cultural Encyclopedia of the 1960s and 1970s*. Berkeley: Ten Speed Press.

McKittrick, David and McVea, David (2002) *Making Sense of the Troubles: The Story of the Conflict in Northern Ireland*. Chicago: New Amsterdam Books.

Mort, Frank (2010) *Capital Affairs: London and the Making of the Permissive Society*. New Haven; London: Yale University Press.

Naish, J. (2008) *Enough*. London: Hodder & Stoughton.

Nicholson, Virginia (2019) *How Was It For You? Women, Sex, Love and Power in the 1960s*. London: Viking.

O'Farrell, J. (2009) *An Utterly Exasperated History of Modern Britain: Or 60 Years of Making the Same Stupid Mistakes as Always*. London: Transworld.

Pellew, Jill and Taylor, Miles (2021) 'Introduction'. In Pellew, Jill and Taylor, Miles (eds.) *Utopian Universities: A Global History of the New Campuses of the 1960s*. London: Bloomsbury. Pp. 1–18.

Riley, Mark and Harvey, David (2007) 'Oral Histories, Farm Practice and Uncovering Meaning in the Countryside'. *Social & Cultural Geography*, 8(3): 391–415.

Rycroft, Simon (2011) *Swinging City: A Cultural Geography of London, 1950–1974*. Farnham; Burlington: Ashgate Publishing.

Sandbrook, Dominic (2005) *Never Had It So Good: A History of Britain From Suez to the Beatles*. London: Abacus.

Sandbrook, Dominic (2013) *Seasons in the Sun: The Battle for Britain, 1974–1979*. London: Penguin Books.

Saunders, Robert (2012) '"Crisis? What Crisis?" Thatcherism and the Seventies'. In Jackson, Ben and Saunders, Robert (eds.) *Making Thatcher's Britain*. Cambridge: Cambridge University Press. Pp. 25–42.

Saville, I. D. and Gardiner, K. L. (1986) 'Stagflation in the UK since 1970: A Model-Based Explanation'. *National Institute Economic Review*, 117(1): pp. 52–69.

Seabrook, Jeremy (1978) *What Went Wrong? Working People and the Ideals of the Labour Movement*. London: Victor Gollancz Ltd.

Shaw, Gareth, Curth, Louise and Alexander, Andrew (2004) 'Selling Self-Service and the Supermarket: The Americanisation of Food Retailing in Britain, 1945–60'. *Business History*, 46(4): 568–582.

Singer, Jefferson A. (2004) 'Narrative Identity and Meaning Making Across the Adult Lifespan: An Introduction'. *Journal of Personality*, 72(3): 437–460.

Sium, Aman and Ritskes, Eric (2013) 'Speaking Truth to Power: Indigenous Storytelling as an Act of Living Resistance'. *Decolonization: Indigeneity, Education & Society*, 2(1): I–X.

Somers, Margaret (1994) 'The Narrative Constitution of Identity: A Relational and Network Approach'. *Theory and Society*, 23(5): 605–649.

Spencer, Colin (1993) *The Heretic's Feast: A History of Vegetarianism*. London: Fourth Estate.

Tönnies, Ferdinand (2012 [1887]) 'Community and Society'. In Lin, Jan and Mele, Christopher (eds.) *The Urban Sociology Reader*. Abingdon: Routledge.

Tyrrell, Naomi and Harmer, Nichola (2015) 'A Good Move? Young People's Comparisons of Rural and Urban Living in Britain'. *Childhood*, 22(4): 551–565.

Valentine, Gill (1997) 'A Safe Place to Grow Up? Parenting, Perceptions of Children's Safety and the Rural Idyll'. *Journal of Rural Studies*, 13(2): 137–148.

Wall, D. (2005) *Babylon and Beyond: The Economics of Anti-Capitalist, Anti-Globalist and Radical Green Movements*. London; Ann Arbor: Pluto Press.

Wandersman, Abraham and Nation, Maury (1998) 'Urban Neighborhoods and Mental Health: Psychological Contributions to Understanding Toxicity, Resilience, and Interventions'. *American Psychologist*, 53(6): 647–646.

2 Making the move

Why Wales?

The stories in Chapter 1 explained why people wanted to leave cities and suburbs. While the stories hinted at why these migrants chose to search for a new life in the countryside, they gave little explanation of why they chose to move to Wales specifically. A desire for the countryside does not automatically equate to a desire to live in Wales. Barbara, Josie, and Judith all mentioned being influenced by friends who had already moved to the area, but what drew the friends there in the first place? To find out what attraction Wales had for the countercultural migrants of the 1960s and 1970s, we need to delve deeper into the stories of those who moved as well as the story of what was happening in Wales itself.

2.1 The ironic appeal of rural depopulation

I will begin this section by sharing another of the migrants' stories, this time that of Sue, from Bath. In 1978, Sue moved to rural mid-Wales in search of a more adventurous lifestyle. Prior to the increase in numbers of migrants during the 1970s, the population of rural Wales had been shrinking, and Sue became immediately aware of the drastic impact this out-migration trend had been having on the area:

> During the seventies here, the place was dying on its feet. Mid-Wales, the whole of mid-Wales was dying on its feet. People with any, you know, people went away to college, they went away to university, they didn't come back. It was exactly the same on the west coast of Ireland. Because there was nothing here. There was no work.

It was something that had been going on for a while. There had long been a trend towards young people moving out of the country in search of employment and more exciting lifestyles; it was something which affected virtually all of rural Britain at this time, but Wales was particularly badly hit. The first Beacham Report, compiled in 1951, showed that the depopulation plaguing the British countryside was at its worst in mid-Wales. Here, just as Sue correctly identified, the decline of traditional industries had led to people leaving in search of new employment (Halfacree and Boyle 1992: 10). Between 1871 and 1951, the area lost 25% of its population, and numbers continued to decline until at least 1971 (Walford 2004: 313).

DOI: 10.4324/9781003358671-4

To some extent, this was the fault of the rapid urbanisation occurring elsewhere. Writing in 1978, P. Commins argued that 'growth in prosperous areas tends to draw resources away from depressed regions. These in turn become poorer, unemployment increases, and migration will lead to gradual depopulation' (Commins 1978: 81). Since one of the characteristic features of the 1960s was the prosperity Britain was experiencing, particularly with regard to manufacturing industries, it could be argued that this was itself drawing people away from rural Wales. This would mean that the very thing which was pushing countercultural enthusiasts out of cities – rising consumerism – was also the thing which was pushing young people out of the rural areas the countercultural migrants would soon end up in.

As we learnt earlier, Barbara had been an out-migrant from rural Ireland, and so she was particularly aware of the relationship between out- and in-migration:

> [T]he village that we came to visit, um, which is where I live now, there were a lot of empty houses. A lot of, it seemed to me that a lot of the young people were leaving Wales, maybe for the same reasons that I'd fled Ireland, I don't know.

Rural depopulation was one of Wales's most significant characteristics in the years leading up to the countercultural migrants' arrival. The effects it had were many and varied, but they had significant implications for the countercultural migrants' choice of location.

When Barbara recalled her first visit to the village where she now lives, she remembered how she 'was really struck by the peace, the quiet, and the emptiness really'. The sense of Wales as empty was one of the side effects of rapid depopulation, and this apparent emptiness made the landscape seem wilder and closer to nature. Away from the obvious industrialisation of the valleys or the deep gouges of the North Wales slate landscape, humankind's mark on the Welsh landscape was hidden from the casual observer. Josie remembered how the letters from a friend in Wales 'just seemed to drip greenery'. Lush, green pastures gave the impression of having been left untouched for generations. This appealed to countercultural migrants' desire to separate themselves from cities and suburbs.

Day et al. observe that accounts of why people liked and chose to move to Wales feature curiously few mentions of local people, and this is reflected in the interviews collected here. Though many speak at length about the locals when they come to share what life in Wales was like, locals rarely appear in accounts of their preconceptions (2008: 112). Depictions of Wales from the time, many of which influenced the decision to move, often shared this characteristic. For example, Elizabeth West's (1977) account of moving to a rural Welsh hill farm spends four full pages giving a description of their home and surroundings, but no other people get a mention until page three, when 'Welsh women' get a cursory acknowledgement in relation to national dress and a reference to a friend from England on page four. Instead, the image painted is one of empty wildness: 'we are so isolated up here that my tattered array of patched sheets, darned woolies, frayed towels and threadbare knickers is viewed only by the odd passing raven' (15–18).

The image of an empty wilderness is still prevalent in contemporary depictions of Wales. This excerpt from journalist Neil Ansell's *Deep Country* was written in 2011, but might just as easily have come from one of the 1970s in-migrants:

> This uninhabited swathe of the Cambrian Mountains right in the very heart of the country has been called the green desert of Wales, its empty quarter.
>
> Just downhill across the track there was once a farmhouse, presumably Penlan Farm. An overgrown rocky mound in the approximate shape of a large building from which now sprout full-grown ash trees, it is not so much a ruin as the ghost of another era. The landscape is far wilder than it once was; the hundred-and-twenty-acre hill farm around me now barely supported a single elderly tenant farmer, yet scattered around its fields were the remains of five farmhouses or labourers' cottages, and a mill.
>
> (Ansell 2011: 3–4)

The apparent emptiness of Wales manifested itself in the form of empty buildings such as these. We will be looking more closely into the homes migrants settled into later on, but for the moment, it is useful to consider the attraction the empty homes provided. When countercultural escapees were looking for somewhere to start a new life, they found Wales to be full of empty houses scattered throughout the landscape. Barbara connected the empty houses to the lack of people, saying that 'In the village that I eventually moved to there were really empty houses and no people really'. At the same time, Josie recognised how this meant there was more scope for people to move in:

> There was a, a lot of people moving down here, mainly buying derelicts and building them up, places that had, you know, because of the move away from the country to the city a lot of the old remote places were falling apart.

Lesley and Ian were also among those who bought up derelicts to refurbish, and even the house Judith moved to had been empty.

Many migrants felt that the plethora of abandoned properties made Wales more affordable, and this gave it much of its attraction. Josie recalled how

> quite a lot of people throughout those early 70s, like the people who were moving here, um, wanted to get back to the land, to move away from the consumerist society, looking for a different way of living. And the land was cheap and the houses were cheap.

Note how the justification here for choosing Wales was not just its distance from consumerism but also its low cost. Moving to Wales was a more manageable permanent escape than the drastic moves to Afghanistan, India, or mainland Europe that some other migrants were making. This argument is repeated in the literature on regional migration and counterurbanism (see, for example, Day et al. 2008; Walford 2001). However, statistics from this era suggest that Wales was not drastically cheaper than other rural areas. Chris Hamnett's (1983) report on regional house prices found that the average cost of a house in Wales was roughly the same as those in the North West, Yorkshire, Humberside, and the East Midlands (102). So why was it that this particular group of migrants thought it was cheaper? There were two main types of houses becoming empty through depopulation: farms and smallholdings that were completely empty and being sold along with some of the original land, and empty buildings which formed part of farms still owned and inhabited by locals and were available to rent from the farm owners, sometimes along with use of a neighbouring field or two. Often, these buildings had never been modernised, having either been abandoned too long ago or being too isolated for it to be worthwhile (again, this is something we will come back to later). I argue that the number of *derelict* properties is what made rural Wales such an affordable choice for countercultural migrants, despite the fact that average house prices there were no lower than much of the rest of Britain. As discussed earlier, Wales suffered more severely from out-migration than other parts of Britain, meaning it had a plethora of abandoned housing which was genuinely cheap. Countercultural migrants were keen to inhabit derelicts because they fit with the ethos of living lightly on the Earth. And on top of this, the idea of empty land ready for the taking, as disturbingly colonialist as it might be, tied in nicely with the aim of getting away from existing social structures.

The availability of inexpensive housing meant that Wales also attracted people who would really have preferred to move elsewhere but could not afford it. Recollections from two migrants new to our story would help to exemplify this. When Mike and his family decided they wanted to take up farming, they had originally wanted to move to the West Country, but

> having travelled from Dorset down to Cornwall, we realised that was out of the question, on price, you know you'd be lucky if you got a house for the money we'd got, even then, in anywhere decent. So we thought, what's the next thing . . .? Somebody suggested Wales.

They ended up purchasing a Carmarthenshire farm in 1977. Likewise, Freya and George were another couple who found themselves working around their finances.

> We had got this idea that we wanted to have some land, but Devon was a far too expensive place, we were living in Devon and we could never have afforded it. And um, so wherever we went we looked around and thought 'can we move here?', like Scotland or Ireland we went to, not on the job but we went sort of 'round and about. And um, Wales was the cheapest place.

Even in Wales, they could only purchase an off-grid property jointly with another couple, but since this fit in with their goals for self-sufficiency, Freya and George moved to Powys in 1975. This was what made Welsh counterurbanisation more affordable – and therefore attractive – than elsewhere: the extent of the depopulation, combined with the length of time it had been going on for, meant that dilapidated properties were more plentiful than in other inexpensive rural areas.

2.2 Revitalisation and popularity

There was a flip side to this story of abandonment and depopulation. Heather, who we have not met before, left Liverpool for Wales in 1976, along with her husband and young son. In many ways, the family fitted the hippie stereotype. In 1973, they attempted to travel the overland route to India but got stuck in Afghanistan for a month due to a military coup.[1] In Liverpool, they lived in a series of squats, moving six times in two years, and their son did not go to school. But when they then decided to try moving to the country, their reasons for choosing Wales as their destination were altogether more conventional:

> We chose mid-Wales because if you were signing-on, you could relocate and the government would give you £500 to relocate. And they would also give you your travelling expenses. But you had to go to a job, you had to have a specific job that you were going to, which would last for six-months.

The mention of this scheme to encourage migration to rural Wales reveals another explanation for the area's affordability. Schemes like this feature in other migrants' reasons for moving to Wales; while their motives for leaving the city may have been idealistic, their motives for choosing Wales specifically were often much more practical.

These kinds of opportunities were created to counter the effects of the out-migration that had been steadily eating away at the population of rural Wales. Following the findings of the 1951 Beacham Report, county councils in mid-Wales responded by setting up the Mid-Wales Industrial Development Association, whose purpose was to encourage economic growth in order to stop people leaving in search of employment. Thanks to the efforts of this and the Mid Wales Development Corporation (and the Development Board for Rural Wales, which replaced them both in 1977), 83 new factories opened in rural Wales between 1957 and 1973, increasing the area's manufacturing employment by 83% (Halfacree and Boyle 1992: 11–14). Sue, who earlier described the extreme depopulation of mid-Wales, eventually settled in Newtown. This was one of the towns most affected by development efforts:

> There was huge redundancies you know, on the Wirral, Liverpool, Manchester area, Birmingham, um, you got a really big influx of people coming to the area because there was, they were offered housing and they were offered jobs. . . . So many people moved into Newtown and it went from a, a mainly Welsh population to um, over two-thirds were English. Practically in ten years.

The man who would go on to become Barbara's husband was one of the people attracted to Wales because of employment, with Barbara recalling that 'he came here for a job, he wanted to get out of London. Um, yeah he wanted to get out of London but only came for the job really'. People frequently chose Wales because of job offers – even hippies had to earn money somehow – and this was one of the biggest draws to the area. Countless migrants arrived to take up employment as university lecturers, veterinarians, teachers, factory employees, engineers, town planners, doctors, and more. Even when searching for an escape from 'the rat race', they still needed to know that they would be able to support themselves and their families, so the availability of suitable employment was a strong factor in dictating where they ended up. A job was a stepping stone to a new life. In 1988, the Institute of Welsh Affairs conducted a survey among migrants to rural Wales, many of whom would have been part of the countercultural group discussed here, and found that employment was one of the key factors in attracting people to the area (alongside the scenery, quality of life, and low crime rates) (Walford 2004: 323).

Plenty of the new arrivals to Wales had never visited beforehand and had few ideas about what to expect. Barbara was one of these people. With little prior experience of the country, she was, as she reflected later, woefully ignorant that Wales would be at all different from England:

> I don't think that I actually realised before I came to Wales that um, there was a Welsh language. That sounds, that sounds shocking now. . . . I didn't really realise that there was such a distinct culture. Um, the only time I'd ever really been in Wales before was coming from Ireland off the boat in Holyhead, through the north of Wales which is all mountains, and down to London.

A similar lack of knowledge was echoed throughout the contributors' stories. Take a new migrant, Maureen, for instance. In 1969, Maureen's husband secured a job at Aberystwyth University, and with three-month-old twins in tow, the family set off for their new life. Maureen, originally from Scotland, had never set foot in Wales before her arrival:

> [T]he only thing I knew about Aberystwyth was that Prince Charles had been here at university, the headlines, but I never thought about it really. I never, I probably never thought that Wales would be different from England. I hate to say that, but I probably didn't.

This all suggests that many people were primarily attracted to Wales because of what it offered them in terms of lifestyle rather than because of an attraction to any inherent Welshness. But at the other end of the spectrum were those who already had prior connections to and experience with the area. For them, the attraction of choosing Wales as their destination came primarily from the idea of returning to and strengthening these connections, rather than economic benefits or visions of wildness. The connections came in a variety of forms. Wales had been a popular holiday destination for many years, and for many, its attraction came from a desire to revisit the site of childhood summers. Ruby moved to Wales in 1982, slightly later than the main batch of migrants, and for her, the holiday memories were coupled with family connections:

> As a child, myself and my sisters spent most of our school holidays with my Nain and Taid and spinster aunt in Dinorwic. . . . I guess the biggest reason

> was quite an overwhelming sense of 'hiraeth' for Wales and a strong desire that the language would continue and that my three children would learn to speak as well.

The phenomenon of people with Welsh connections choosing to return to the country is perhaps another side effect of the out-migration it has experienced over the preceding decades. These returnees would likely be, like Ruby, the children and grandchildren of those who had been driven out to the cities. The connections could sometimes be distant, but they were enough to pull people back in the direction of Wales.

Of course, once a few people had moved to Wales, word spread, and the number of migrants snowballed. The significance of this phenomenon is not to be underestimated, and even in cases where separate draws to Wales can be identified, the initial suggestion of Wales as a possibility often came from friends who had already moved. Barbara, Josie, and Judith were all influenced by friends who had already made the move to Wales. Humans are herd animals; we tend to go where we know others.

Nigel Walford has suggested that there are two main models of explanation for the move from depopulation to growth in rural Wales: structural and behavioural. The structural model suggests the growth was caused by increasing employment opportunities and industrial redistribution. The behavioural model suggests that people were attracted to the countryside because of the way they perceived rural life (2004: 312). Stories from those who moved show that not only were both models factors in the decision but also that they weren't as separate as they first appeared. Rural depopulation contributed to the image of Wales as an empty wilderness. Depopulation also led to the availability of affordable (albeit derelict) homes, which again helped to strengthen the image of wildness. At the same time, attempts to halt and reverse depopulation provided economic incentives and employment opportunities for migrants. The very same phenomenon which made Wales seem wild also meant that Wales was seeing increased investment and development. It worked: between 1971 and 1981, the population of mid-Wales increased by 7.4%, and the majority of this growth was due to in-migration by people attracted to rural areas. The growth figure for Wales as a whole was only 1.8% (Halfacree and Boyle 1992: 26). So there were many reasons why people were attracted to Wales, some more conventional than others. This twin narrative of conventional vs. alternative reasons continues through the following chapter.

Note

1 This was the coup led by Daud Khan against his cousin, King Zahir Shah, which ultimately ended the monarchy and established Afghanistan as a republic (Barfield 2010: 155).

References

Ansell, Neil (2011) *Deep Country: Five Years in the Welsh Hills*. London: Penguin.

Barfield, Thomas (2010) *Afghanistan: A Cultural and Political History*. Princeton; Oxford: Princeton University Press.

Commins, P. (1978) 'Socio-Economic Adjustments to Rural Depopulation'. *Regional Studies*, 12(1): 79–94.

Day, Graham, Drakakis-Smith, Angela and Davis, Howard H. (2008) 'Migrating to North Wales: The "English" Experience'. *Contemporary Wales*, 21(1): 101–129.

Halfacree, Keith and Boyle, Paul (1992) 'Population Migration Within, Into and Out of Wales in the Late Twentieth Century: 1. A General Overview of the Literature'. Migration Unit Research Paper 1.

Hamnett, Chris (1983) 'Regional Variations in House Prices and House Price Inflation 1969–81'. *Area*, 15(2): 97–109.

Walford, Nigel (2001) 'Reconstructing the Small Area Geography of Mid-Wales for an Analysis of Population Change 1961–95'. *International Journal of Population Geography*, 7: 311–338.

Walford, Nigel (2004) 'Searching for a Residential Resting Place: Population In-Migration and Circulation in Mid-Wales'. *Population, Space and Place*, 10: 311–329.

West, Elizabeth (1977) *Hovel in the Hills: An Account of 'The Simple Life'*. London: Faber and Faber.

3 Finding a home

Mainstream or alternative?

Once the decision to move to Wales had been made, the first task was usually to find somewhere to live. The significance of this stage is easy to overlook since it occurs so early in the stories of those who moved, but it is important not to underestimate it. This is one of the few aspects of the experience which is universally shared by all migrants. They all had different motivations and aspirations, but nobody can move to Wales *without* finding somewhere to live. This chapter outlines the different decision-making processes migrants went through in order to find a house and the factors that influenced these decisions.

Moving to a new house is one of the most stressful experiences a householder can go through. The pressures of moving have been found to affect women particularly badly, making them especially relevant here given that the majority of stories collected here come from women (Magdol 2002: 558). Aside from the general stress of sorting out the house purchase, packing belongings, figuring out a new locality, etc., moving into a different house involves (unless it is an entirely new build) 'entering the unknown space which previously housed the most secret and potentially shocking aspects of a previous occupants' life' (Chapman and Hockey 1999: 12). Because of this, it was through the experience of choosing and arriving at their new home that migrants became fully immersed in their new life for the first time, at the point when things began to seem painfully real. The mental image they had built of Wales and their expectations for life there could begin to crumble as soon as they arrived.

There are a range of factors that led to the migrants choosing the homes that they did. These create two 'routes' into Wales and the counterculture: one shaped by the influence of urban countercultures and the wider housing crisis, the other created by popular trends in housing and Do It Yourself (DIY) culture. Most of the migrants' experiences fit into these routes, which is how I came up with them, but we should also be careful to remember that not everyone fits neatly into categories in this manner. Some will have been influenced by elements of both routes, some by only part of a route, and some by neither.

3.1 The rebellious route to self-sufficiency

Given that the availability of housing was what drew many to Wales in the first place, the process of finding a house could often be surprisingly difficult. As discussed

DOI: 10.4324/9781003358671-5

earlier, the draw of cheap housing eventually worked against the migrants, as the increased popularity of the area pushed prices up and out of budget. Lesley was one of the migrants who found themselves unexpectedly affected by these rising house prices. When we last saw Lesley, she and her partner were heading off to do the overland trip to India, then they intended to buy a smallholding in Wales. They had already purchased the goats they planned to raise, who were boarding with a friend. But when Lesley and her partner returned from India, and they found the situation in Wales was no longer what they had expected. She remembered how

> it was the time of the oil crisis and three-day week, everything cost a lot more than planned. Plus there were so many people wanting to move to the countryside that the little places which were dirt cheap suddenly became not dirt cheap.

The couple went on a camping trip to hunt for affordable houses and eventually found one, though it was distinctly less than the smallholding they had hoped for: 'we ended up just with a, a cottage, an unmodernised cottage with about quarter of an acre, and a mattress and trying to wheedle my way into grazing stuff for the goats'. The cottage had in fact been partially modernised at one point, but it still lacked plumbing or effective heating. Access was also an issue, and Lesley recalled how on their very first day, 'our removal van got stuck down by the house because we didn't have a track, so that was an interesting introduction, sort of getting hauled out by the local farmer'. Another problem was that the small amount of modernisation which had been done did not correspond with the way Lesley and her partner wanted to live. They therefore had to put a lot of work into making the cottage and land into something that would work for them, and they lived with very little in the way of facilities in the meantime. All the while, the focus was on making a home that would fit with the kind of lifestyle they were attempting to achieve:

> Sanitation took a long time, um, proper heating so it didn't just have an open fire. We had to take out these horrible brown tiled fireplaces that seemed to be in every single place and putting in a wood burning stove. 'Cause CAT [the Centre for Alternative Technology] opened about that time as well, so we learnt about wood burning stoves and things like that. Yeah, and insulating the house, there was no insulation. And my parents were very into conservation so I knew back then about sort of insulation, and the importance of stuff like that. . . . We were trying to get it back to an old cottage, but in a nice way.

Jo, whose story we have not been introduced to yet, was another incomer who found herself living in more rustic accommodations than anticipated. In 1975, Jo and her husband were newly married. She was studying nursing in London while he was at university in Stirling. But Jo had an unconventional upbringing, living in New Zealand for five years, and so she had developed a taste for the outdoors. Because of this, she wanted to move to a small town in a rural area. It happened that her husband was unhappy with the degree at Stirling and decided a transfer to a different university would be the best thing. Aberystwyth offered him a place, and this was how the couple ended up moving to Wales. They arrived with no housing organised, and since they needed something quickly, they moved between temporary properties while searching for somewhere to buy. The experience of living in temporary accommodation was quite an ordeal, and certainly not the idyllic introduction to the countryside Jo had hoped for. First came an 'absolute slum' in Aberystwyth: 'we had one room in a really . . . house of multiple occupancy that was basically disgusting'. After that, was a borrowed summer home in Borth:

> The house in Borth was draughty and freezing cold. I remember going to bed in all my clothes and then, I suppose I didn't kind of get dressed in the morning but if I didn't go to bed in all my clothes I would take all my clothes into bed in the morning to get dressed under the bedclothes, it was so cold.

None of this was helped by the fact that Jo and her husband had arrived in Wales for the winter of 1975–76. January 1976 saw Britain hit by the worst storm in 23 years, as sea surges occurred across the west coast and 24 people were killed as a result of the strong gales (Loader 1976: 273). These were hardly ideal conditions for settling in Wales, and certainly not by the coast. The advantage of living in such grim conditions was that Jo and her husband were able to save and, eventually, find and buy a house in the countryside with help from Jo's parents. But even with this financial assistance, the best they could get was a derelict property 'we found a little cottage eight miles from Aberystwyth, by a stream, absolutely idyllic, but not much facilities or anything, so it was quite low because of those days, lots of those places were just falling into ruin'. Jo and her husband moved into their new home just in time for the famously hot, dry summer of 1976, a stark contrast to the winter storms. Because of this, their new home was a far more pleasant experience than the cold, damp temporary accommodation:

> I thought 'oh, this is good! This is alright!', 'cause it was by a little stream that we did washing, and did our washing up in, and everything was really attractive even though it was so basic. I don't think the basicness had an impact on us until the winter came along.

But of course, when winter came, it hit hard. This only served to illustrate the extent of the work which needed to be done to the cottage:

> And then it was quite, quite rough, the snow came in and. . . . The fireplace, it was an old old range, we just put sticks in and the heat when straight up the chimney, so we hung up a heavy curtain halfway across the room, to make the space smaller that we were trying to heat. Curtain went straight up the chimney, because there was such a lot of wind!

Plumbing, heating, and electricity – these were all things that Jo and her husband had to install and/or renovate in order to bring their idyllic cottage up to scratch.

Jo had wanted an outdoorsy life in her idyllic cottage by the stream, just as Lesley wanted a rustic, peasant lifestyle, but for both of them, the standard of property they could afford seems to have been lower than they were expecting.

The image of young people moving into derelict, unmodernised hovels is one of the most prevalent stereotypes of the countercultural movement and a characteristic feature of many of the stories shared. It makes a compelling narrative to focus on the hardships of going without, struggling against the elements, and working to achieve a better life. But there are two main flaws with this narrative. Firstly, as I have just pointed out, Lesley and Jo had not initially intended to live in ruined cottages. They wanted a peasant lifestyle, but nothing in their stories suggests they planned to go quite that far. And secondly, because of the inspirational nature of the flawed narrative, it is easy to forget that the experience of living in outdated housing was not restricted solely to the rural counterculture movement. Plenty of people, not just hippies in rural Wales, were dealing with poor living conditions in the 1960s and 1970s. The baseline of what was considered acceptable housing was so much lower than today that while it may seem that the migrants to Wales were dealing with exceptionally hard conditions, they were probably not much worse than some of them had experienced before moving. For example, the average indoor temperature for UK homes in 1970 was 13.7°C, more than 3°C below the modern average of 16.9°C. In the 1960s, central heating was installed in just 5% of UK households, and coal fires continued to dominate until well into the decade; by 1971, the proportion of homes with central heating was still only 30% in England and 36% in Wales (Kuijer and Watson 2017: 77–78; Townsend 1979: 480–481). Even in homes that did enjoy the luxury of central heating, the efficiency of the technology was only around 65%, a far cry from the 90% efficiency we enjoy today (Roberts 2008: 4483–4484).

Heating is just one example of how the standards of housing in the 1960s and 1970s differed from today. But of course, not all housing reached the expected standard to begin with. The Housing Act of 1969 declared that all homes should have: 'an internal WC, fixed bath or shower, wash-basin, hot and cold water at three points and a sink'. An estimated 25% of English and Welsh homes failed to meet these criteria (Townsend 1979: 480). These substandard dwellings were not restricted to rural areas. In London, for example, the 1969 Greater London Development Plan stipulated that an exclusive water supply (hot and cold), bath, and indoor toilet were essential amenities, and dwellings lacking these facilities were to be deemed unfit. Only 36% of dwellings in Hackney met these requirements (Wall 2017: 80–81). The homes purchased by incomers such as Lesley and Jo may have been at the bottom end of the spectrum of 1970s housing, but they were certainly not unique. It may seem like a digression, but the wider state of British housing in the 1960s and 1970s is relevant to the story of the incomers and the homes they ended up in – in several ways. It is not just a simplistic point about how houses in the 1970s were worse than people remember and everyone was hard up; that is just the first part of the story. The poor state of housing in the 1960s and 1970s and the post-war attempts to redevelop Britain with contemporary, modernist architecture had a knock-on effect on the types of homes people ended up in when they arrived in Wales and how they chose to renovate them.

Britain in the 1960s and 1970s was suffering from a housing shortage. It was a problem that had been going on for a while, and even as home ownership became more attainable for some, thanks to rising wages and higher disposable incomes, a significant number of people were unable to find somewhere to live (Langhamer 2005: 343). This was partly what post-war rebuilding and slum clearance attempts were trying to resolve. But the processes of clearing old houses and building new ones didn't always match up properly, meaning that the shortfall sometimes worsened precisely because of attempts to remove it. In 1976, there was a shortage of 300,000–400,000 houses in Britain, but in the 1971 census, only a few years earlier, there were found to be over 675,000 dwellings left empty (Marshland 2018: 30; Milligan 2016: 8–9). While some were second homes or kept empty for some other purpose, a vast number were buildings deemed not to meet the new housing standard, as discussed earlier, and consequently awaiting demolition. In Greater London alone, 23,100 houses were waiting to be demolished in 1971 (Wall 2017: 79–81). As the number of homes available dropped, the cost of those remaining rose rapidly:

> House prices rose by around 60 percent between 1958 and 1963. A clear sign of the growing shortage was the fall in the differential between the price of new and of existing houses, which had stood at around 30 percent in 1959 but had virtually disappeared by 1963. In the inner urban area much of this older property was of poor quality, lacking standard amenities, and subject to overcrowding levels which hastened deterioration.
>
> (Davis 2001: 72)

Lesley's story showed that by the time she and her partner returned from travelling the hippie trail, they could no longer afford to purchase anything more than an isolated off-grid cottage. Jo was also restricted to near-ruins, even with financial help from her parents. These were not isolated incidents or experiences restricted to the fringes of society; they were part of a story told by thousands of people in these years. This was the high price that came with modernisation; inevitably, not everyone could benefit immediately.

Sue was one of the thousands of people unable to find a home. The last we heard of her, she was making the decision to leave her home in Bath and move to Wales, attracted by its sparse population and the potential for a more adventurous lifestyle. Before she moved to Bath, Sue had lived in London, and there she had seen first-hand, the extent of the problems caused by poorly planned redevelopment:

> They built this, the most hideous flats you have ever seen in your life. And of course within months of these flats opening they just turned into rat nests. Because the crime rate in them just went through the roof, you know. And nobody wanted to live there, and eventually, and I think they've knocked them down since, you know. And that's forty years ago. And they did that to a whole huge swathe of houses, and they were doing it all over London. And people were getting really beed off about it, and that's why, yeah, that's why

> the big squat scene started in London. Because of that, you know, and plus the fact that I mean, you know, a basement flat which was usually the cheapest was probably three-quarters of your wages a week. You know, you barely had enough to live on.

Sue herself lived in one of these London squats. Although her local councillor and neighbours were supportive of their plight, she and her fellow squatters faced a significant amount of bullying and harassment from the police and other authorities. This was in itself one of the reasons for moving out of the country.

> [I]t was very corrupt, the 1970s. So if you were kicking against the traces, it was, it got, it got very difficult in London, and it got very difficult in the cities. You know, I think a lot of people left because of it.

But the struggle to get by in the city was more than just a reason for moving to the country; it was also one of the reasons why people ended up in such poor housing once they got there. Heather (introduced earlier) was another migrant to Wales who had first lived in a squat, this time in Manchester:

> Spent two years and we lived, just for two and a half years, something like that, and we lived during that time in six different places, because we moved from squat to squat. Because again it was really hard to find accommodation then, there was still a lot of areas which were um, well they weren't bombsites but they appeared to be bombsites. Where they'd pulled the old housing down but they hadn't built anything.

As discussed earlier, the only reason that Heather was able to move to Wales was because of the financial incentive provided to people willing to relocate. Like Jo, Lesley, and Sue, she had to save for a bit first, and then the house she ended up in was in poor condition and needed refurbishing: 'we pitched up here and um, rented, and then found a very cheap house which we then renovated for years on the edge of Llandrindod'.

There were also official squats. A character new to our story, Millie, inhabited one of these before she moved to Wales:

> We lived in a flat in Haverstock Hill, in London, that's Camden. It was SCH, which was Student Community Housing, so it was like a legalised squat because Student Community Housing had gone to Camden Council and asked them to allow people to live in the houses that were going to be demolished or were going to have something else happen to them. So you got a three month house, or a six month house, if you were really lucky you got a two year house. And actually, I have to say, this was all back to Jeremy Corbyn, who at the time was part of SCH. There was him, and his brother, and his younger brother, and he was the man who actually went to the council and said in a nice way, can we please use some houses?

Since this was student housing organised in collaboration with Camden Council, it was a more formal set-up than Sue's, and Millie's account of her time squatting is far more positive in nature. But even then, squatting was at best a short-term solution – as Millie says, only the lucky ones got a two-year house – and moving out of the city was often the only way of getting a home of one's own. Millie and her husband were offered a free holiday in Wales with friends, and while there they fell in love with the area and decided to move immediately, initially living in a static caravan provided by the farm where they found jobs picking potatoes. It then took financial assistance from both sets of parents before they were able to buy a small bungalow.

Officially understood as the uninvited occupation of empty buildings, the legal status of squatting in the UK is complex, and in the 1970s it depended heavily on a variety of factors, including the type of building, precisely how the squatters had taken residence, the extent of the owner's knowledge and involvement, and so on (Finchett-Maddock 2014: 215). Since squatters did (and still do) have some rights under UK law, squatting was frequently seen as a way of protesting against the combined lack of housing and plethora of available buildings. It is a way of enacting social change: 'to occupy or counter-occupy, in other words, is to insist on building the necessary conditions for social justice and new autonomous forms of collective life' (Vasudevan 2015: 318). Numerous countercultural groups formed squatting communities; there was the Brixton Gay Community, for example, and the feminist, women-only squat in Hackney (Cook 2013: 85; Wall 2017: 80). Estimates for the number of squatters vary, but it is generally believed that there were around 40,000 by the mid-1970s (O'Hara 2006: 288; Marshland 2018: 36). But while squatting may have originated as, and certainly been perceived as, a way of making a political statement, this was far from always the case. In 1977, the Self Help Housing Resource Library conducted a survey of London squatters, the results of which were published in *Squatters: Myth and Fact*, which showed that 96% of respondents were only squatting because they were unable to find affordable housing (Milligan 2016: 13). John Marshland records how 'officials even admitted in private that they had to tolerate squatting because they could not envision an alternative' (2018: 44).

So the squatters that came to Wales to live in derelict cottages were not just making a statement about sustainability; they were working with what they could afford. I named this 'the rebellious route' because, at first glance, it seems like the more politically charged, countercultural reason for moving into near-ruins. But what it actually shows is that the kinds of homes that the countercultural migrants to Wales purchased were not only reflective of the values they hoped to embody through their new lifestyle. Indeed, they could sometimes be the opposite, as in the case of Lesley's poorly modernised house. Instead, these homes reflected the wider, ongoing struggle for housing.

3.2 The mainstream route to self-sufficiency

There was a second, more mainstream reason why people were so keen to buy run-down houses once they had arrived in rural Wales. This was also connected to the housing crisis and improvements in expected living standards, but in a more indirect way. Because they so frequently ended up in homes of a primitive nature, many migrants expected that they would have to bring their new properties up to a liveable standard themselves. Most people, like Jo, had little practical experience with renovation work before moving and learnt as they went along. Ian (introduced earlier) fell into this group. When he turned 21 and had finished at university, Ian inherited a small sum of money from his mother and purchased a derelict smallholding for £4,500:

> It was very difficult because I had no practical skills really, so I was sort of learning on the job, and it did need a lot doing. Um, but we did it up quite a lot and I lived there for some time. . . . I'd been looking for some time and that was about the only one I could afford.

Once upon a time, the skills needed to renovate an entire smallholding would have entailed a great deal of specialist knowledge. Crafts like masonry and carpentry were passed down through the generations, and aspiring craftsmen completed lengthy apprenticeships to reach the required standard. But by the 1960s and 1970s, these skills were under threat from the increased mass production of goods and in danger of dying out entirely. This had two consequences. Firstly, it meant that some groups felt a growing pressure to preserve and maintain these traditional skills in perpetuity. Secondly, it opened up the possibility of making them easier by mass producing some of the elements needed and allowing people to finish the process themselves (without enduring years of apprenticeships first). And so the DIY industry was born.

The factors that contributed to the sudden growth of DIY link back to the events discussed earlier. One of these was home ownership. Part of the explanation for the housing crisis of the 1960s and 1970s was an earlier spike in home ownership in Britain, as those fortunate enough to benefit from the affluence of the 1960s made use of their newfound wealth. In 1945, owners occupied only 26% of houses in England and Wales, but by 1966, this figure had grown to 47% (Nixon 2017: 77). The rate of growth was so rapid that Britain went from having the lowest proportion of owner-occupiers in Europe in 1948 to the highest proportion in 1985 (Martens 1985: 607). As home ownership increased, people began to see the potential of DIY to improve their homes and add extra value to them, something they would not have been able to do in rented properties. DIY offered a chance to realise aspirations of something better for new homeowners who could not afford to hire tradesmen; it was a way of asserting control and ownership (Powell 2009: 91–92). This was an activity very different from the crafts of the old. People were engaged in DIY as a lifestyle choice; it was a leisure activity or a way of raising their status and standard of living, not a necessity (Atkinson 2006: 2). It was in the 1960s and 1970s that shops dedicated to selling goods for the DIY market, such as B&Q, first emerged. Prior to this point, raw materials could only be purchased at builders' merchants and other specialist outlets, but by 1983, the major DIY chains owned more than 300 stores between them (Watson and Shove 2008: 73; Jones 1984: 65–67).

DIY was a necessary skill for countercultural migrants. When Jo found herself in that idyllic stream-side cottage, she and her husband were faced with a great deal of work to do in order to bring the property up to standard. Electricity had to be connected, plumbing installed, and heating updated, as well as all the general structural and cosmetic adjustments that moving into an older property entails: '[It was] really difficult few years, and when my first daughter was born we still didn't have a toilet, bathroom, proper kitchen or anything, but by the time the second daughter came along we did. . . . It was like bliss!' Paul Atkinson argues that along with the growth of DIY came a 'de-skilling of the processes involved'; the growth of instructional books and DIY shops meant that one could get started more easily (Atkinson 2006: 5). In this respect, the counterculture movement was a key part of the growing trend in DIY. One significant link was the publication of the *Whole Earth Catalogue* in America. First compiled by Stewart Brand in 1968, it contained articles on DIY techniques and products (Wakkary et al. 2015: 611). Wakkary et al. argue that the publication of the *Whole Earth Catalogue* was a watershed moment for the development of DIY and did much to help strengthen the movement. They recount how 'Steward Brand saw in print the power to bring together "counter-culture communities, back-to-the-land households, and innovators in the fields of technology, design, and architecture"' (ibid: 611). Similar publications appeared in the UK, including *Alternative London* (Saunders 1970) and *Alternative England and Wales* (Saunders 1975). Together, these offered a unique compilation of alternative businesses and organisations, providing everything that one would need to put DIY ambitions into practice (as well as much more besides). These publications show how DIY was not just a necessary skill for countercultural migrants; it was something actively created and promoted by the counterculture movement itself.

As homes became less affordable during the 1970s, interest in DIY only increased. And as interest levels rose, DIY moved further and further into the mainstream. Design historian Deborah Sugg Ryan recounts how:

> At the beginning of the 1970s, the general shortage of spare cash and the rationing of mortgages by building societies meant that many people who wanted to could not afford to move house. The Daily Mail Ideal Home Exhibition responded to this by showing how to get a better home when money was tight.
>
> (Ryan 1997: 145)

Ryan shows that the Daily Mail Ideal Home Exhibition is surprisingly relevant to the story of countercultural migration. While it may seem like the very antithesis of a movement rebelling against consumerism, the growing enthusiasm for DIY provides a link between the two. Exhibitions in the 1970s included a Ministry of Housing stand on renovating older properties, of particular relevance to those buying tumble-down shacks in Wales. In 1975, a special display helped raise awareness of recycling and pollution by showing how materials could be repurposed. 'Rooms for Waste' displayed examples of furniture and storage solutions created from a wide array of materials, including old tin cans, unravelled cardigans, flip-top cigarette packets, milk crates, and drainpipes (Ryan 1997: 145 & 154). Most significantly of all, at the 1974 Ideal Home Exhibition, John Seymour had a stall entitled 'The Fat of the Land' (Sandbrook 2013: 3–4).

John Seymour (1914–2004) was one of the most significant figures in the rural countercultural movement. His books on self-sufficiency inspired countless people to change their lifestyles and provided guidance on what to do once they had. Described as 'the apostle of self-sufficiency', John Seymour published a grand total of 41 books on travel, self-sufficiency, and rural life, as well as presenting numerous television and radio documentaries (Conford 2008: 221; Girardet 2004). He became something of a legendary personality in the field (no pun intended), and his appearance at the Ideal Home Exhibition suggests that this prominence had spread beyond the constraints of the counterculture movement. This was, after all, a mainstream event.

Seymour's 1976 work *The Complete Book of Self-Sufficiency* was the bible relied on by many. When Freya and George (introduced earlier) arrived in Wales, they were completely clueless as far as the actual skills required to run a smallholding were concerned. Freya's sister, Gillian, who moved a couple of years later, was in the same situation. The two recalled:

Gillian: . . . learning how to milk a cow, learning to de-ball a lamb or whatever. We knew nothing.

Freya: . . . We had Seymour's book in one hand and a knife and a pig in the other.

And it was not just animal husbandry that Seymour taught. When Freya and George arrived at their off-grid property in Powys, there was not only no electricity but no water either:

Freya: [It was] a ruin so it was very cheap and it fitted our budget . . . it was a total change. It was basically like camping . . . you know like washing the nappies in the stream by hand and um, that sort of, it was basic. And now you'd never do it, but then things were different.

Lighting came in the form of oil lamps. With most of the group preoccupied with childcare and earning money to support their new lifestyle, it fell primarily to George to work on improving their home. First came the installation of gas: 'we went up the valley to a house which was, now had a generator and electric and bought their calor gas fittings so we had gas lights'. After this came a 12V windmill. This was followed by a hydropower generator on the small stream which ran through their land. Hydropower did not work in the summer when the stream was too small or in the winter when it was frozen, so solar panels were also fitted. Water was hand-pumped to save on energy use. These were all topics covered by John Seymour. *The Complete Book of Self-Sufficiency* gave many the confidence to believe that they could put a self-sufficient lifestyle into practice. It, along with other DIY manuals of the time, provided easy access to knowledge that would have been extremely difficult to obtain only a few years earlier. And as this example shows, it was not just old knowledge of traditional crafts that Seymour was passing on, but also cutting-edge information on alternative energy sources.

As well as giving countercultural migrants the confidence to believe that they could survive in an off-grid hovel in rural Wales, books like this impacted directly on the kind of homes they ended up in by giving property advice. In *The Complete Book of Self-Sufficiency*, Seymour provided guidance for laying out one-acre and five-acre smallholdings, with no suggestion that anything smaller might be suitable (Seymour 1976: 19). His earlier book, *Self-Sufficiency: The Science and Art of Producing and Preserving Your Own Food* (co-authored with his then-wife Sally), states that 'a family with four children could live very well on five acres of good land, buying very little from outside, but only if they managed their affairs carefully' (Seymour and Seymour 1973: 15). In keeping with the DIY trend, he writes on the assumption that much needs doing to prepare your new property; there is information on land clearance and drainage as well as insulation and natural power sources (Seymour 1976: 19–26). The focus is on practicality. In his earlier account of self-sufficient life, *The Fat of the Land*, he openly admits that he has little skill in creating a comfortable house:

> [T]he Seymours are not really nature's house dwellers. We are really gypsies, or nomads, or rovers. . . . Our house is apt to look like somewhere where we are camping for the moment; and this is the antithesis of what a peasant family's house should be.
>
> (Seymour 1961: 59)

Other authors, perhaps inspired by Seymour's success, also penned guides to running smallholdings and rural life in general. They too tend to give rustic images of the domestic side of rural life. C.J. Munroe's (1979) volume *The Smallholder's Guide* dedicates more space to choosing a property with good outbuildings than to considerations regarding the house itself, where he simply states that: 'there are additional things to be looked at – water pipes, wiring, the efficiency of water heating, the cooking stove'. Further consideration of the cooking stove suggests that a solid fuel stove like an Esse or Aga would be better and cheaper than installing gas or electricity (28). With advice like this, it is no surprise that so many people bought houses in poor condition.

One of these people was Jim. In 1979, Jim was a firefighter living in Swanley, Kent, with his wife and two small children. Although he had faced a difficult upbringing in Tottenham's infamous Broadwater Farm Estate – one of the post-war developments that seemed to exemplify urban life in the era – Jim had managed to escape a life of inner-city poverty and crime. With a good job and a nice suburban house, he and his family initially seem unlikely candidates for dropping everything and moving to Wales. But as Jim recalled:

> Everyone was into sort of self-sufficiency and the good life, it was just in the atmosphere I think, you know. But we, trying to think back to what might be of interest to you, I mean there were two, the two books that were very . . .

> pivotal, I think, in that movement was I Bought a Mountain, I don't know if you know that, by [Thomas] Firbank. It's the story of a couple who bought a hill farm in North Wales. That's me, and my wife as well, it's a story, so we read that and it was a sort of really idealised, the dream sort of thing. Although, but then it all went pear shaped. . . . And the other one was John Seymour.

It was because of these books' influence that Jim and his family uprooted to Wales, and the reason why they ended up buying: 'a very small smallholding in Wales which was basically a single room hovel with a thatched roof with corrugated tin on top'. Choosing a nice house was not a priority for those inspired by this movement. The burgeoning DIY culture was telling people that houses could be renovated, upgraded, and personalised. They did not need something perfectly formed right at the beginning. Although, as it happened, Jim did end up paying builders to construct a new house on the land:

> We lived in a caravan, an old static caravan for six months while we had the house built, which I had no idea what I was doing, anyway I just got builders in to do it and then it got spent and I had some money to pay them, so . . . it was all harum-scarum in that sense.

From this, it may seem that the countercultural migrants to Wales were not so far removed from the consumerism of conventional society after all. They were part of a wider cultural shift of aspiring towards homeownership and home improvement. People were keen to improve their living standards, even when they were only able to purchase properties that were old, rundown, and unmodernised. Countercultural migrants were unusual because they tended to move farther than a lot of people, often to rural locations. Wanda and David moved to North Wales in 1979 after falling in love with the area while on holiday there. Wanda recalls how:

> We lived in London because I was a special needs teacher and I had um, I had the ability to get a flat at a reasonable rate near where I was working. So that seemed like a good starting point since we didn't have any money. Saved like billy-o for a year, and got the North Wales Weekly every week on a Saturday, on Saturday, and ticked the houses where we might like to live.

They too had been influenced by John Seymour and other books in the same vein, selling candles and other homemade crafts at a stall in Putney:

> I don't know where it was going. And we went, oh we were doing things like painting our own pictures and putting them like, I don't expect you remember Athena Prints but there used to be, Athena Prints was a big shop where they sold blockboard basically, with prints stuck onto the surface. You had a dry mounting press so we were putting our own pictures on blockboard and selling them, and it was like 'oh, we could make a living doing this.'

Their story illustrates how these kinds of activities were situated at the crossroads of the conventional and the countercultural. Wanda reflected that 'we're still relatively conventional in the sense that the first thing we did is buy a house. And we did only do that after having, having jobs'. It's an astute observation which applies to many of those who moved. The situation was different for people who joined communes or made homes for themselves in the stereotypical yurts and static caravans that have come to dominate images of the counterculture, but they were a small minority. The majority of the incomers were people who ended up in tiny, rundown cottages through a combination of it being all they could afford and influence from John Seymour and the self-sufficiency and DIY movements. John Benson has suggested that the counterculture:

> modified – rather than totally rejected – traditional patterns of consumption, spending a good deal of what money they had on the purchase of products such as music and drugs, rather than fashion and furniture. Thus hippies too used consumption – and non-consumption – as a means of defining and expressing their particular view of the world.
>
> (Benson 1994: 166)

The growth of the DIY industry ensured migrants could easily access the materials and knowledge they needed. There was certainly a steep learning curve, and life in these ramshackle cottages was far from easy, but it is interesting that we saw so much less countercultural migration to Wales before DIY took off. Rather than concocting an entirely new trend themselves, the counterculture was simply taking an existing trend to a new extreme.

So far, throughout these three chapters, I have sought to find out why so many people chose to move to rural Wales in the 1960s and 1970s. The answer to this question is far from simple. The reasons for countercultural migrants' arrival in Wales are a complex web of interlinking factors. For some, it may have been primarily due to the perceived unpleasantness of city life, exacerbated by the impacts of the oil crises, strikes, IRA threats, modern housing developments, and so on. Or maybe it was not so much cities they disliked, but the increase in commercialisation and capitalism and the loss of a strong sense of community. For others, it was perhaps due to the attractions of country life combined with the appeal of purchasing an abandoned property to do up, spurred on by the new range of DIY products and guides. Then again, maybe it was just a lifelong dream, something not linked to any clear factors but simply present: a personal wish. Whatever the reason for their move, there is a clear pattern becoming apparent here: a mixing of conventional and alternative cultures and ideas. People are not confined to one or the other, not in their motivations for moving or in their choice of dwelling. What appears on the surface to be the most countercultural of choices (like moving to Wales) may in fact be impacted by more conventional influences as well (like the financial incentives available). The counterculture was not, in this case at least, a cult which had to be entered into completely; mixing ideas seems to have been both acceptable and commonplace.

References

Atkinson, Paul (2006) 'Do It Yourself: Democracy and Design'. *Journal of Design History*, 19(1): 1–10.

Benson, John (1994) *The Rise of Consumer Society in Britain, 1880–1980*. London; New York: Longman.

Chapman, Tony and Hockey, Jenny (1999) 'The Ideal Home as It Is Imagined and as It Is Lived'. In Chapman, Tony and Hockey, Jenny (eds.) *Ideal Homes? Social Change and Domestic Life*. London: Routledge. Pp. 1–14.

Conford, P. (2008) '"Somewhere Quite Different": The Seventies Generation of Organic Activists and Their Context'. *Rural History*, 19: 217–234.

Cook, Matt (2013) '"Gay Times": Identity, Locality, Memory, and the Brixton Squats in 1970s London'. *Twentieth Century British History*, 24(1): 84–109.

Davis, John (2001) 'Rents and Race in 1960s London: New Light on Rachmanism'. *Twentieth Century British History*, 12(1): 69–92.

Finchett-Maddock, Lucy (2014) 'Squatting in London: Squatters' Rights and Legal Movement(s)'. In Van Der Steen, Bart, Katzeff, Ask and Van Hoogenhuijze, Leendert (eds.) *The City Is Ours: Squatting and Autonomous Movements in Europe from the 1970s to the Present*. Oakland: PM Press. Pp. 207–232.

Girardet, Herbert (2004) 'John Seymour: An Ecological Pioneer Championing the Cause of Living Simply'. *The Guardian*. Available at: www.theguardian.com/society/2004/sep/21/environment.guardianobituaries. Accessed 9 September 2019.

Jones, Peter (1984) 'The Retailing of DIY and Home Improvement Products'. *Service Industries Journal*, 4(1): 64–70.

Kuijer, Lenneke and Watson, Matt (2017) '"That's When We Started Using the Living Room": Lessons from a Local History of Domestic Heating in the United Kingdom'. *Energy Research & Social Science*, 28: 77–85.

Langhamer, Claire (2005) 'The Meanings of Home in Postwar Britain'. *Journal of Contemporary History*, 40(2): 341–362.

Loader, C. (1976) 'The Storm of 2–3 January, 1976'. *Journal of Meteorology*, 1(9): 273–304.

Magdol, Lynn (2002) 'Is Moving Gendered? The Effects of Residential Mobility on the Psychological Well-Being of Men and Women'. *Sex Roles*, 47(11/12): 553–560.

Marshland, John (2018) 'Squatting: The Fight for Decent Shelter, 1970s–1980s'. *Britain and the World*, 11(1): 27–50.

Martens, M. (1985) 'Owner-Occupied Housing in Europe: Postwar Developments and Current Dilemmas'. *Environment and Planning A*, 17: 605–624.

Milligan, Rowan Tallis (2016) 'The Politics of the Crowbar: Squatting in London, 1968–1977'. *Anarchist Studies*, 24(2): 8–32.

Munroe, C. J. (1979) *The Smallholder's Guide*. London: David & Charles.

Nixon, Sean (2017) 'Life in the Kitchen: Television Advertising, the Housewife and Domestic Modernity in Britain, 1955–1969'. *Contemporary British History*, 31(1): 69–90.

O'Hara, Glen (2006) 'Social Democratic Space: The Politics of Building in "Golden Age" Britain, c. 1950–1973'. *Architecture Review Quarterly*, 10(3/4): 285–290.

Powell, Helen (2009) 'Time, Television, and the Decline of DIY'. *Home Cultures*, 6(1): 89–107.

Roberts, Simon (2008) 'Altering Existing Buildings in the UK'. *Energy Policy*, 36: 4482–4486.

Ryan, Deborah Sugg (1997) *The Ideal Home Through the 20th Century*. London: Hazar Publishing.

Sandbrook, Dominic (2013) *Seasons in the Sun: The Battle for Britain, 1974–1979*. London: Penguin Books.

Saunders, Nicholas (1970) *Alternative London*. London.
Saunders, Nicholas (1975) *Alternative England and Wales*. London.
Seymour, John (1961) *The Fat of the Land*. London: Faber and Faber.
Seymour, John (1976) *The Complete Book of Self-Sufficiency*. London: Faber and Faber.
Seymour, John and Seymour, Sally (1973) *Self Sufficiency: The Science and Art of Producing and Preserving Your Own Food*. London: Faber and Faber.
Townsend, Peter (1979) *Poverty in the United Kingdom: A Survey of Household Resources and Standards of Living*. Harmondsworth: Penguin.
Vasudevan, Alexander (2015) 'The Autonomous City: Towards a Critical Geography of Occupation'. *Progress in Human Geography*, 39(3): 316–337.
Wakkary, Ron, Schilling, Markus Lorenz, Dalton, Matthew A., Hauser, Sabrina, Desjardins, Audrey, Zhang, Xiao and Lin, Henry W. J. (2015) 'Tutorial Authorship and Hybrid Designers: The Joy (and Frustration) of DIY Tutorials'. *CHI*: 609–618.
Wall, Christine (2017) 'Sisterhood and Squatting in the 1970s: Feminism, Housing and Urban Change in Hackney'. *History Workshop Journal*, 83: 79–97.
Watson, Matthew and Shove, Elizabeth (2008) 'Product, Competence, Project and Practice: DIY and the dynamics of craft consumption'. *Journal of Consumer Culture*, 8(1): 69–89.

Part 2
Practices

Once they had arrived and settled in Wales, the only thing left for the migrants to do was carry on with their lives there. The following three chapters look at how living an 'alternative' life plays out in practice. Each centres on a different aspect of everyday life: self-sufficiency, making money, and time away from work. By pulling apart these separate strands of the migrant experience, we can get a fuller picture of exactly what their lives were like and how they connect to the wider countercultural ethos and events of the time. Again, it has to be remembered that there is not going to be a clear distinction in everyone's lives between the practices of self-sufficiency and their way of making money, or between time working and time away from work. This is something I acknowledge throughout these chapters.

It is through understanding the day-to-day lives and activities of these migrants that we can begin to understand how they enacted their countercultural identities. Understanding identity has long been a concern of human geography, from Paul Vidal de la Blache's ideas of identities as an integral part of geographic regions to the further development of this theory by humanist geographers in the mid-twentieth century and through to the present (Seamon and Lundberg 2017). Ideas of 'otherness' and identity recur throughout these three chapters, apparent mainly in the division between 'mainstream' and 'countercultural' identities and activities. The whole idea of a '*counter*culture' requires something to actively counter, as Elizabeth Nelson points out in her work on the topic (1989: 1). Identity has historically been understood as heavily dictated by the differences between people, constructed around the idea of 'self' vs. 'other' (Said 1978). Stuart Hall argued that 'identities can function as points of identification and attachment only because of their capacity to exclude, to leave out, to render "outside", abjected. Every identity has at its "margin" an excess, something more' (1996: 5). In dividing identities into boxes or categories, it's possible to forget the complexities that underlie identity in reality. I try to negate this by acknowledging the margins and the identities which exist outside of boxes. There are so many intertwining relationships at play here that it is almost impossible not to consider the way they blur together. As we will see, these migrants' everyday lives demonstrate how they constructed blended identities through their activities.

DOI: 10.4324/9781003358671-6

The centrality of everyday experiences to constructions of identity is a core principle of these chapters. This idea emerged as part of feminist approaches to geography in the 1970s, with Charles Lemert arguing that this is because

> the women's experience is (for better or worse) rooted in practical interests of daily life to a far greater extent than men's experiences which, in the 1970s especially, was based on their working in the relations of ruling in government and corporate life where objective, official knowledge was primary.
>
> (2011: 23–24)

This vision of identity as rooted in everyday life would help pave the way for a series of reimaginings over the next few decades – reimaginings that would help shape identity into the concept I use to understand the migrants' everyday lives. Gillian Rose, in particular, drew parallels between feminist theory and human geography, highlighting that 'identity is relational. Who I think I am depends on me establishing in what ways I am different from, or similar to, someone else. We position ourselves in relation to others' (1993: 5; see also, Gilligan 1982; McDowell 2013, 2016). The idea of identity as relational is important for understanding how migrants used their lifestyle choices to separate themselves from mainstream culture.

bell hooks's writing has also influenced the ideas in these chapters, especially regarding the implications of identity as experienced over time. Given the scope of this project, which encompasses the whole of the contributors' life stories (or at least as much as they wish to share), this is particularly useful. Drawing on postcolonial theories to understand identity, hooks writes that:

> [S]ince decolonisation as a political process is always a struggle to define ourselves in and beyond the act of resistance to domination, we are always in the process of both remembering the past even as we create new ways to imagine and make the future.
>
> (2015: 4–5)

This view of identity as a continuing struggle between different factors, shifting through time, is extremely valuable. In this way, we start to see identity becoming more than just 'self' vs. 'other', more than simple binaries of inclusion/exclusion. This is something that will be referred to in the following chapters, as we see how the migrants' combined elements of 'mainstream' and 'counter' cultures in their lives.

In the 1990s, a branch of research developed that dealt with the ways identity is used to include and exclude, particularly in relation to place. Writing in 1995, David Sibley commented that the social sciences seemed to be becoming better equipped to tackle problematic ideas around social exclusion, thanks to developments in feminist literature, critical race theory, and so on. However, he acknowledged that it is impossible to rectify the gap completely without engaging with people who have been 'othered' in some way, writing that 'I see the question of making human geography radical and emancipatory partly as a question of getting close to other people, making way for them' (184). While utilising the idea of object-relations theory, Sibley argued that 'exclusion' can only be understood properly if we also think closely about 'the self' or 'the way in which individual identity relates to social, cultural and spatial contexts' (ibid: 4). This ties Sibley's notion of self into the ideas discussed earlier.

Paul Cloke and Jo Little's edited collection *Contested Countryside Cultures* (2005) dealt with many of the same ideas as Sibley's but from a specifically rural perspective. They argued that 'one of the potential hazards of a focus on "otherness" is that research will privilege particular forms of other to the neglect of other others' (10). Reinforcing a narrow view of rurality through research can act as an exclusionary device, continuing to keep marginalised groups from identifying with and being identified as part of rurality. Cloke and Little write that 'as rurality is increasingly understood as a phenomenon which is socially and culturally constructed, so the exclusionary qualities within these constructions need to be highlighted' (ibid: 4). As rural society continued to diversify and change in character, these practices of inclusion/exclusion continued to be of interest, especially when it comes to understanding the way different identities are perceived and treated in rural contexts (see Neal 2009). The following chapters build on all of these pre-existing ideas by placing the stories of these migrants in their historical surroundings while ultimately focusing on their lives and experiences.

References

Cloke, Paul and Little, Jo (2005) 'Introduction: Other Countrysides?'. In Cloke, Paul and Little, Jo (eds.) *Contested Countryside Cultures: Otherness, Marginalisation and Rurality*. Abingdon: Routledge. Pp. 1–17.

Gilligan, Carol (1982) *In a Different Voice: Psychological Theory and Women's Development*. Cambridge, MA: Harvard University Press.

Hall, Stuart (1996) 'Introduction: Who Needs "Identity"?'. In Hall, Stuart and du Gay, Paul (eds.) *Questions of Cultural Identity*. London: SAGE. Pp. 1–17.

hooks, bell (2015) *Black Looks: Race and Representation*. New York: Routledge.

Lemert, Charles (2011) 'A History of Identity: The Riddle at the Heart of the Mystery of Life'. In Elliott, Anthony (ed.) *Routledge Handbook of Identity Studies*. London: Routledge. Pp. 3–29.

McDowell, Linda (2013) *Working Lives: Gender, Migration and Employment in Britain, 1945–2007*. Chichester: Wiley-Blackwell.

McDowell, Linda (2016) *Migrant Women's Voices: Talking about Life and Work in the UK since 1945*. London: Bloomsbury.

Neal, Sarah (2009) *Rural Identities: Ethnicity and Community in the Contemporary English Countryside*. Abingdon: Routledge.

Nelson, Elizabeth (1989) *The British Counter-Culture, 1966–73: A Study of the Underground Press*. Basingstoke: Macmillan.

Rose, Gillian (1993) *Feminism and Geography: The Limits of Geographical Knowledge*. Cambridge: Polity Press.

Said, Edward W. (1978) *Orientalism*. New York: Random House.

Seamon, David and Lundberg, Adam (2017) 'Humanistic Geography'. In Douglas Richardson (ed.) *The International Encyclopedia of Geography: People, the Earth, Environment and Technology*. Chichester: Wiley-Blackwell.
Sibley, David (1995) *Geographies of Exclusion: Society and Difference in the West*. Abingdon: Routledge.

4 Living the dream? The reality of self-sufficiency

For many countercultural migrants, the realities of being 'alternative' meant facing the practicalities of living a self-sufficient lifestyle. Buying a derelict smallholding was only the first stage of the process; once they moved in, they still had to deal with the day-to-day reality of this life. Although not all countercultural migrants were attempting self-sufficiency, this was the experience of a significant proportion of them. Even for those who were not dedicated to the idea, moving into remote, derelict housing without electricity or plumbing meant experiencing some elements of self-sufficiency, whether they wanted to or not. To get a better idea of what this was like, this chapter is broken down into three themes, each focusing on a different aspect of life for self-sufficient migrants:

1 The rhythms of time and daily life, looking at their day-to-day activities and how they compared to conventional life at the time.
2 The significance of crafting and practical skills to this lifestyle.
3 The particular problems and experiences faced by women in coping with the realities of being female in a traditionally masculine environment.

While it is impossible to cover every dimension of this group's lives with the level of detail they deserve, I hope that by focusing on these three topics I'll be able to give an overview of their experiences.

4.1 Spending time, saving money

To understand the practicalities of living a self-sufficient lifestyle, we begin by exploring the rhythms of daily life on smallholdings. Theories of how to understand 'everyday life' in a conceptual context vary tremendously (Sandywell 2004: 161; Kalekin-Fishman 2013: 726). Here I take the concept literally, and my discussion of everyday life focuses on the different types of activity that filled the migrants' days. This requires a certain amount of extrapolation, as none of the interviews I conducted actually contained a blow-by-blow account of a typical day (perhaps because, as anyone who works with animals and the land will attest, there is no such thing as a 'typical day' on a smallholding). Nevertheless, by looking at the different activities they do mention, we can get some idea of the mundane, often unspoken, tasks that go along with them.

DOI: 10.4324/9781003358671-7

Take George, Freya, and Gillian's story, for instance. Between them, they kept two flocks of sheep, along with dairy cows, pigs, chickens, and bees. Animal husbandry, and all the regular tasks that go along with it, will therefore have taken up a great deal of their time. In the short term, cows must be milked, chickens fed, pigs mucked out, and so on, plus annual tasks such as shearing, lambing, and slaughtering (this last task always proved a challenge; Freya recalled how she could only eat the chickens they killed after storing them in the freezer for long enough to 'become anonymous'). Their move into farming involved a steep learning curve, which meant that along with looking after their animals, a significant amount of time had to be taken up with figuring out how to care for them in the first place. As an example, Freya and George originally planned to have a house cow – one in milk but with no calf, which means she must be milked twice daily – and to breed pigs from a gilt they had purchased. They would then be able to fatten the piglets using the excess milk. But although the theory seemed sound, it never quite played out that way:

Freya: The reality proved a lot different, the pig never got pregnant she had something internal wrong with her. And the cow died in fact and um, that was just terrible. And um, it was a long time, lots of vets and illnesses, much loved cow. Um, so, and the pig had to go as well, and ultimately we ended up with a series of house cows which we kept the calf, which was the idea we got from them.

Even something as basic as choosing what breed of cow to keep was surprisingly complicated. Jerseys and Guernseys produce the best milk, but Dexters or Welsh Blacks are better suited to the land. Likewise, the pigs proved to be a perpetual source of trouble, thanks to their tendency to eat absolutely anything, from children's welly boots to young lambs:

Gillian: You'd see her running around the fields with the ewes that had their lambs with them, perfectly normal ones, she'd be running and galloping, she was huge, galloping around the field with her mouth open chasing a lamb. I mean she wouldn't catch it of course.

The knowledge of how to deal with these problems had to be learnt, and then put into practice. Days became filled with trial and error, as well as grass and mud. They were not alone in this, as many of the new self-sufficiency enthusiasts were also learning on the job – one interviewee bought a dairy farm when the closest he had ever been to a cow previously was looking through the car window.

Along with animals come animal products, and processing these added another load of work. George and Freya's and Gillian's households both made their own cheese, as well as churning their own butter. The sheep provided wool as well as meat, and this was all hand-spun into yarn (even though none of them knew how to knit; in the end, Freya used some of it for crochet as well as for that most 1970s of crafts: macramé). The skins from slaughtered animals were cured at home, which could lead to some interesting interactions:

Gillian: I'll never forget [our daughter] had a, I think she was at uni, anyway had a friend who'd never been to our house before, who said 'I can't understand your parents' house, they've got dead sheep laying around everywhere'. Which we did, didn't we?

They found themselves, especially George, developing a huge range of practical skills, from carpentry to stonework, electrics to plumbing; even the spinning wheel was homemade. This process of constantly learning and creating was immensely satisfying, as George recalled: 'I enjoyed doing practical stuff. . . . I got feedback from it, satisfaction. If something works, it's great'. Of course, it was not just caring for animals, processing animal products, and DIY tasks that occupied the two households' time: no self-respecting smallholding would be complete without a vegetable garden. George, Freya, and Gillian grew as much as possible in theirs:

Freya: . . . everything you can really.
Gillian: Ranging from peaches and apricots and lemons, in [Freya's] case, to leeks and carrots.
Freya: Everything. Because originally, what you didn't grow, you didn't have.

So weeding, watering, and feeding were piled onto the days' schedules alongside everything else. And as with caring for animals, these regular tasks were accompanied by annual ones: sowing, harvesting, pruning, and cutting – all of which must be completed at the appropriate time of year or risk jeopardising the garden (Duckham 1963: 392).

You may have noticed that all of these tasks are, to some extent, dependent on and affected by the seasons. Some of these connections are obvious, as in the case of the vegetable harvest or lambing season. Others are less apparent. Cow's milk, for instance, varies throughout the year according to what she is eating and the stage of growth her calf is (or would be) in. Some months it will be better for hard cheese, others for soft cheese or butter, and so on (Chen et al. 2014: 216). Large animals should be slaughtered in the autumn or winter, not just to reduce the need for winter stores of animal feed and to provide meat, but also to avoid the problems caused by summer flies (Seymour 1976: 106). This means that the processes of preserving the meat and drying the skins take on a specific season in turn. Likewise, sheep are always sheared at the same time of year (usually midsummer, though it varies across the UK), which means that the jobs of processing, spinning, and using the resulting wool take on a seasonal rhythm (ibid: 120). DIY tasks can also vary according to the weather – projects requiring outside space cannot be done in the cold or wet, for instance – and the availability of materials, which, as with wool, fluctuate through the year (Jones 1964: 21). This means that what the self-sufficient family did each day was dictated to a much greater extent by the time of year, and not just the current weather conditions but the past and future conditions as well.

Take this example from E.L. Jones's (1964) book on agricultural history, *Seasons and Prices*:

> Where the rhythm of the seasons is influential, the uncertainties of the weather will be crucial, for the wind and rain alert the yield of all the most trivial harvests of the countryside, and shift forwards or back from one year to another the dates on which they may be gathered. They determine whether or not the countryman shall have jam on his bread and butter, how much jam, and which fruits shall go into its making, and how much effort has been put into acquiring it.
>
> (1964: 28–29)

The town dweller may, in extreme cases, face higher prices for jam if the season has been bad, but that's the worst-case scenario. Successful self-sufficiency relies on harnessing the power and resources of the natural world and working with elements of the environment that can't be changed. This requires a detailed knowledge of when and how things grow and an awareness of what each season requires (Spence 2009: 6). The everyday lives of the countercultural migrants who attempted self-sufficiency were impacted by seasons in a way that those living 'normal lives' simply were not.

To understand the implications of this more fully, we need to take a step backwards for a moment. Western civilisation has a long history of viewing itself as separate from and more important than the natural world (Winston 1997: vii–viii). Since the Industrial Revolution, as modern lifestyles have become more and more technologically advanced, a division has emerged and deepened between humans and the environment. This division is both physical and psychological, as people have shifted from direct to indirect experiences of nature (Hinds and Sparks 2008: 109). By the 1960s and 1970s, the majority of urban dwellers were experiencing an extreme disconnect from nature (Turner et al. 2004: 588; Pyle 2003: 206).

Living a self-sufficient lifestyle rectified this disconnect. Being more heavily impacted by seasons and weather meant that the countercultural migrants in search of self-sufficiency were also more aware of it. Rain in the forecast takes on more importance when it could mean the difference between a failed harvest and a successful one, rather than just whether to wear a raincoat. And living in off-grid properties, often, as we have seen, without running water or central heating, means inclement weather has more practical implications than it might otherwise. Archaeologist and anthropologist Christopher Tilley observes that: 'Unlike virtually any other artefact in contemporary Western culture, [the garden] is *commonly* acknowledged to have multi-dimensional sensory qualities that overlap and are intertwined and affect the gardener' (2006: 312, emphasis in original). It is not that life on a smallholding is a more sensory experience than life anywhere else – an office can involve just as many smells, sounds, and textures as a farm, though admittedly of a different sort – but that the senses take on an extra level of importance and receive more recognition as a result. Life on the land involves using the senses in a way that is rarely acknowledged in contemporary life, as weather and, by extension, the seasons can only be felt, not touched in the traditional sense. Tim Ingold, also an anthropologist, explains that:

> To feel the wind is not to make external, tactile contact with our surroundings but to mingle with them. In this mingling, as we live and breathe, the wind, light, and moisture of the sky bind with the substances of the earth in the continual forging of a way through the tangle of life-lines that comprise the land.
> (2007: S19)

The use of one's senses, the passing of time, and awareness of the environment all took on new significance when pursuing self-sufficiency. Writing in 1983, John Cherrington began his book of the farming year as follows:

> I was not born into farming, but I chose it as a career because farmers always seemed to me to have an independence from the regimentation of ordinary life, to be able to do as they like when they liked. . . . It was an approach based entirely on a profound ignorance of the real facts of farming; of having every operation dependent on the weather.
> (7)

Cherrington started his adventure in farming some years before the back-to-the-land movement took off, but I think many countercultural migrants would have related to his words. This exploration of day-to-day life on self-sufficient small-holdings has shown that the tasks and activities that filled each day were intrinsically linked to the seasons and the weather in a way that normal everyday life for people in this era was not. Not only was it intrinsically linked, but it also involved engaging with the natural world on a level that people do not usually think about to the same extent. These were days filled with a plethora of sounds, smells, tastes, and textures, all linking together to determine what tasks need doing, and when, and how, and what doing them now will mean for tomorrow.

4.2 A different kind of knowledge: (re)learning old skills

Dependency on the rhythms of the seasons was a significant characteristic of the day-to-day activities of self-sufficient countercultural migrants, but it was not the only significant characteristic. These day-to-day activities were also heavily focused on practical skills of the kind which had once been central to everyday life but were rapidly becoming obsolete. As George, Freya, and Gillian's story showed, the two households were participating in a wide range of crafts – carpentry, crochet, spinning, dairying, and so on – which were all old skills experiencing a resurgence in popularity. Craft historian Andrea Peach argues that, like so many other elements of the counterculture, this resurgence of interest and participation in crafts can be linked to the political and economic turmoil of the 1960s and 1970s. The desire for escapism that led to people moving to the countryside also inspired this nostalgic interest in crafting (2013: 168–169). This section expands further on the day-to-day lives of self-sufficient countercultural migrants to learn more about the role crafts played in the counterculture movement. It was one of the biggest differences between the countercultural lifestyle and mainstream culture at the time, but the reasons behind this distinction are not always as straightforward as they appear.

Corynne, who we have not met until now, moved to Wales from Bristol in 1976. Although Corynne had grown up in a town, she spent regular holidays out at her uncle and aunt's farm, and this experience of the countryside instilled in her a great love of rural life. Unable to stand living in an urban environment surrounded by people and keen for her children to experience the same freedoms she had, Corynne and her family moved to a run-down house in rural Wales. As part of the group of migrants who were inspired by John Seymour and the DIY movement, their new home required a hefty amount of renovation work. And while it could be expected that this would only be a temporary state of affairs, they found themselves moving to a different house in need of renovation nine years later so that they could have more land for the animals and vegetables which formed the focus of their new lives.

For Corynne, the self-sufficiency ethos was closely linked to the 'make do and mend' spirit she had inherited from her mother and which bled through into all aspects of daily life. Her children all wore handmade dresses and dungarees crafted from cheap, market-bought fabric, and any knitwear was made by hand as well. The move to Wales meant that money was in short supply, and this made making do all the more necessary:

> We were quite happy to sort of muddle through, make do, if we hadn't got it in the fridge or, a neighbour gave me an old ice-cream freezer so I could freeze a glut of apples that somebody happened to have. So if I didn't have it in the freezer we went without, you know. There was a little shop in the village, but couldn't really afford to buy much there. But, um, I would do one shop a month in town, make my own bread, and when we moved to where we are now, not only did I bake my own bread, we had a cow, hand-milked her, made my own butter.

In many ways, the life Corynne and her family led was a relic of a bygone age – her stories seem more reminiscent of the 1930s and 1940s than the 1970s and 1980s. And that was exactly what she had come to Wales in search of: 'It was the old way of living, that Wales was still living, and you couldn't do it in England'. This old fashioned, make do and mend mindset was significant. George, Freya, and Gillian discussed it too:

Freya: Part of what we were about was we didn't have any money, so . . .
Gillian: You had time, but you didn't have money.
Freya: So you never stayed in a hotel, you went camping, you know holidays . . .
Gillian: You didn't eat out, that's the other thing . . .
Freya: No, you never ate out.
George: If something breaks, you mend it . . .

Freya: Right. Etc, etc . . . or you buy it second hand from the charity shop. You don't just buy. . . . But beyond that, I mean you could say that we never had the money to do that, but now we've got more money, relatively, we still don't want to do that. It goes against the grain, and it feeds into like recycling, you know, it's totally, it's the same . . .
George: It was our upbringing really, post war upbringing.
Freya: Post war, things were scarce.
George: Make do and mend.

These three refer back to the post-war 'make do and mend' attitude in the same way that Corynne did. It was this kind of nostalgic, idealised image of the past that the stereotypical self-sufficient lifestyle attempted to recreate. On a practical, day-to-day level, this meant spending time making and relearning how to make things which other people were now buying. It all linked back to one of the key tenets of the countercultural movement: the rejection of consumerism. There was a desire to stop the process of working to earn money to buy things, which would only be used up, thus necessitating more working to earn more money to buy more things, and so on. Instead, people wanted to cut the middle process of working at unrelated jobs to earn money and focus their time on creating the things they needed for themselves (see, for instance, West 1977: 18–20; Seymour and Seymour 1973: 8–11).

Subverting the cycle in this way meant going against the overarching attitudes of society at the time. Practical skills, concerned with crafting, creating, and making, have long been regarded by Western society as less worthy or valuable than theoretical, abstract knowledge. And while practical skills were once essential despite this, as time wore on, they became replaced by machines or relegated to underpaid overseas workers. This meant they were less essential to everyday life and even less valued than before. Sociologist Michael E. Gardiner links this change to the loss of seasonal rhythms discussed earlier, arguing that:

> Modernity is generally characterised as a form of social organisation wherein human needs and actions become increasingly subordinated to the technical requirements of a rapidly expanding and centralised apparatus of commodity production, distribution and consumption, instead of being rooted in the more 'organic' rhythms and textures of daily life in the premodern world. Within 'learned' discourse, everyday knowledges became the target of ridicule and vilification.
>
> (2006: 206)

Likewise, in exploring the concept of *cræft* archaeologist Alex Langlands argues that 'we appear to have created a society that looks disparagingly on people who use their hands to earn a living' (2017: 22). This is exactly the process and mindset that countercultural migrants were rebelling against, consciously or not. Their lifestyle choice meant reprioritising a set of everyday knowledges that were being widely disregarded elsewhere.

The 1960s and 1970s saw a continued increase in mass production, consumption, and mechanisation – particularly of food and clothes – as Britain slowly recovered from the Second World War. By the late 1950s, British clothing had become dominated by a small group of large-scale manufacturers (Ewing and Mackrell 1992: 204). Between 1970 and 1976, the portion of the UK clothing market being imported from abroad doubled from 11% to 22%, with the majority of these being cheap Asian products. Employment in the British clothing industry subsequently dropped, from more than 400,000 employees in 1978 to only around 260,000 by 1985 (Gibbs 1987: 313–314). Reduced prices and higher wages meant that young people were now purchasing significantly more than previous generations (Majima 2007: 507). And with clothing becoming cheaper and Britain's population becoming further removed from the realities of its production, it should come as no surprise that this is when the term 'throwaway society' emerged (Hellmann and Luedicke 2018: 83). As clothing production was moving further and further away and becoming more and more mechanised, food production was also becoming more industrialised. In the 1950s, Europe had still been experiencing wartime food shortages, but by the 1970s, improvements in farming had led to surplus harvests across the continent (Evans et al. 2013: 14–15). Technological advances affected the processing of food as well as its production. Sociologist Richard Sennett describes how, between 1970 and 2000, most of the work that went into industrial baking changed almost beyond recognition due to mechanisation. This did not just alter the experience of baking itself but also the roles of employees, as those higher up became skilled in machinery, programming, and management rather than in the actual craft of baking itself (2008: 249). Similar processes occurred throughout the manufacturing and food production industries.

We can see from Corynne, George, Freya, and Gillian's stories that the kinds of practical crafts seeing a resurgence in the countercultural movement can be roughly grouped into three categories: DIY (discussed more fully in the previous chapter); textiles, which cover spinning, sewing, knitting, and so on; and food, which includes baking, dairying, etc. These rough groupings of activities formed an important part of the image of countercultural self-sufficiency promoted by the books and guides detailing the lifestyle. Elizabeth West was one of the earlier countercultural migrants to Wales, arriving in 1965. Her 1977 memoir of life in rural Wales, *Hovel in the Hills*, is one of the lesser-known books that inspired others to move to the countryside in search of 'the simple life'. The way she describes her life corresponds with these categories:

> On hot sunny days you will find us dressed in shorts, working outside. . . . On cold wet days you will probably find us in the kitchen. Depending on the time of day and year, you will find me washing up, making bread, ironing, bottling fruit, boiling up jam on the Primus or watching Alan at work. He will be sitting cross-legged on the floor surrounded by parts of a bicycle, clock, gun, electric fire, branding iron, adding machine, or whatever the current thing brought in for repair happens to be. Or he may be smoking his pipe and watching me at work.
> (17–18)

The bulk of Seymour's *The Complete Book of Self-Sufficiency* is made up of advice relating to food, with chapters on 'Food from the Fields', 'Food from Animals', 'Food from the Garden', and 'Food from the Wild'. These cover various different methods of preparing and preserving food as well as growing it, with recipes given for everything from Norwegian flat bread to parsnip wine. In addition to this, there is advice on various more crafty pursuits, such as how to make bricks, spin cotton, or sew leather (1976).

I do not think the parallel between what was being increasingly mass-produced and what was being made at home by countercultural enthusiasts was a coincidence. While the dominant narrative is that practical skills become less valuable as they fall out of use, there is also a sense of them becoming more nostalgic. As methods of production evolve, those who retain memory and knowledge of traditional techniques become, as sociologist Vasiliki Galani-Moutafi puts it, 'living embodiments of collective natural and cultural histories' (2013: 104). It was not just the guides to self-sufficiency which were creating this new image of crafting in the 1960s and 1970s. In 1974, the Crafts Advisory Council (CAC) was founded. This was a government-backed organisation focused on developing the craft movement in Britain, which offered grants and loans to artists, commissioned work, and provided patronage, as well as organising exhibitions and publicity. In its first three years of existence, it provided over £140,000 worth of funding to individuals and organisations (Peach 2013: 163). There is, however, one key difference between the crafts practised by self-sufficiency enthusiasts and the kind of thing being preserved by the CAC. The CAC was concerned more with 'fine art' than anything with a practical purpose, a branch of making which has consistently been valued above the practical craft skills employed by countercultural enthusiasts (ibid; Mason 2005: 262). For self-sufficiency enthusiasts, on the other hand, crafting was an essential rather than a hobby. As Gillian remarked earlier, 'You had time, but you didn't have money'. Subverting the cycle of working and buying meant that practical skills had to be used on a daily basis out of necessity.

Rejecting consumerism meant that self-sufficient countercultural migrants were spending their time making the things they needed rather than buying them, engaging with the practical skills and materials necessary to survive. This meant using skills which had fallen out of use with increased industrialisation, going against the mainstream trend of using more supposedly valuable skills wherever possible. But it also ties into the second, parallel narrative of growing nostalgia for these 'lost crafts', situated in an idealised image of the past. The crucial difference between the nostalgic view of crafting and the skills employed by members of the counterculture is one of choice vs. necessity. While living a self-sufficient lifestyle was itself a choice, making that choice made crafts a necessity. Opting out of crafts would have meant opting out of the lifestyle. Of course, plenty of people did pick and choose, and we will hear more about them later. But for those with little alternative income – time but no money, as Gillian said – crafts were as much a part of everyday life as the changing of the seasons.

4.3 'A lady farmer': the role of women

> 'When I grow up', said Sophie at breakfast time, 'I'm going to be a farmer'.
> 'You can't', said the twins.
> 'Why not?'
> 'Farmers are men', they said.
> 'Well', said Sophie, 'I'm going to be a lady farmer. So there'.
>
> (King-Smith 1988: 21)

Although it was published a little after our group of migrants arrived in Wales, this excerpt from the 1980s children's book *Sophie's Snail* encapsulates the third key element of the lives of those who pursued self-sufficiency: the role of gender stereotypes. Of particular importance are the challenges faced by women who chose to try and live a self-sufficient lifestyle. As mentioned in the methodology, the majority of people I interviewed for this project were women. Life history interviews are an effective way of understanding the challenges faced by women, as they give contributors control over their own stories, valuing their voices (Riley 2009: 669). The data creates an image of self-sufficiency that is overwhelmingly viewed through the lens of the female experience. Their stories reveal the specific trials, tribulations, stereotypes, and expectations associated with women in small-scale agriculture.

This section takes a closer look at these stories, focusing in on the experiences of three women: Jessica, who we have not met yet; Lesley, introduced in Chapter 1; and Corynne, introduced earlier. Each story helps to highlight a different aspect of being a woman in a rural, agricultural community and the way this affected their experience of everyday life. Gender is one of the basic frameworks through which people interpret the world, telling them what to expect from different people and situations (Ridgeway 2009: 145). Patriarchal gender relations and ideas of female domesticity are the keys to the image of the 'rural idyll', which was such a central part of the countercultural migrants' attraction to Wales (Little and Austin 1996: 103; Hughes 1997: 172). Despite this, it is something which has been largely neglected in research on countercultural migration (Agg and Phillips 1998: 252–253). But gender is just as much of a constant presence in the lives of self-sufficient countercultural migrants as the weather, the seasons, or the rediscovery of old crafts, so it is only right to recognise it alongside them.

Jessica moved to Wales in 1977. She was not directly part of the countercultural self-sufficiency movement, but her story contains many overlaps. Having trained at the Royal Veterinary College in London, she arrived in Wales after securing a job at the Milk Marketing Board's artificial insemination centre in Ruthin. Later, she moved to Aberystwyth, taking up a new post at the Veterinary Laboratories System there. Naturally, both positions placed Jessica within the local agricultural communities in much the same way as the countercultural migrants were. And while not self-sufficient, she kept substantial vegetable gardens: 'I had a rhubarb patch, raspberry canes and I used to grow good cauliflowers. Cauliflowers, cabbage, sprouts. Broad beans, runner beans, those sorts of things'. Although being a female veterinarian in the 1970s could not have been easy, it was in her agricultural pursuits that Jessica really began to come into conflict with the expectations placed on women in rural communities:

> Some of the um, the neighbours, the male neighbours thought this was quite interesting, 'cause their wives didn't dig the garden and so on. So they all assumed that I was in need of a lot of advice. Which they would offer quite unsolicited! . . . I can remember two guys standing on the other side of the road watching me change a wheel on my car, when I came out and found there was a puncture. This was obviously the topic of some comment, in the road.

Rural communities often have an extremely masculinist culture, and by participating in conventionally 'masculine' activities such as digging and mechanics, Jessica did not conform to expectations (Woods 2011: 180). The fact that she was unmarried only served to reinforce this. Men are overwhelmingly the public face of farming and agriculture. Their contributions are seen as more valued; they are the ones highlighted in magazines and other media. Women's position, by contrast, is often hidden and devalued (Morris and Evans 2001: 376; Shortall 2014: 68, 1992: 431 & 440). For a long time, even research into gender and farming focused on masculinity, neglecting the position of women (Little 2002: 667). Welsh culture in particular has a strong tendency to view men as 'farmers', with women relegated to a position as 'helpers' (Price 2010: 92). With no man to 'help', Jessica was about as far removed from the expectations of rural womanhood as possible.

Rejecting the expectations of rural womanhood is a common theme in the stories of countercultural migrants. Feminism was a key part of the 1960s and 1970s, so it may be that rejecting expectations was simply something these women were doing anyway. I argue that there are also more complex reasons for the difference. One of the reasons why farming is seen as a masculine domain is because of the convention of passing farms down through the male line, thus giving a patrilineal family history of connection to that specific plot of land (Woods 2011: 219). Lise Saugeres has argued that, in the West, this means that men have come to be considered 'natural' farmers with an inherent connection to the landscape, passed down through generations. Women lack this connection since they are excluded from the familial succession (2002: 373). Gloria Leckie explains the significance of this phenomenon:

> land is a valuable commodity, a source of power and wealth. Traditionally, the rural patriarchy has maintained a position of strength by ensuring that both the knowledge base and the land base of agricultural production have been passed primarily to other males.
>
> (1996: 310)

Countercultural self-sufficiency enthusiasts, who were essentially buying their way into the lifestyle through purchasing smallholdings, were not part of this culture of patrilineal inheritance. Not having been brought up with these expectations, they found it easier to disregard them. And besides, they were finding their way around a new set of cultural expectations and often lacked the expected prior knowledge, experience, or inclination required to do things 'properly'. This meant that many, like Jessica, found themselves circumventing the traditional way of doing things.

Lesley was another migrant who took a less conventional approach to life on a smallholding. She had far more agricultural experience than her journalist husband, having worked for the National Institute of Agricultural Botany in Cambridge before the move to Wales. This meant that the practical farming tasks naturally fell to her. As with Jessica, Lesley faced initial scepticism from the neighbouring male farmer, but she recalled how

> when he found that I'd actually been to horticultural college and I could drive a tractor and all these things, within a week I was helping out on the farm. . . . And that was really fortunate, he didn't have any children and I sort of became honorary farmer's son.

Gender roles did not just shift because of prior knowledge but also because of the specific type of agricultural life self-sufficiency enthusiasts were after. Operating farm machinery, such as by driving tractors, was traditionally regarded as a man's job. This meant that as farming became increasingly mechanised, more and more of the new tasks of operating this machinery fell to male farmers, leaving women with a lack of responsibility and a set of redundant skills (Brandth 2002: 188). But for this group of smallholders, there was much more emphasis on living an old-fashioned sort of rural life, with a rejection of technology and machinery. The focus was not on profit, as was the case on most farms by this point, but simply on subsistence. And with less work delegated to machines, there was less work being reassigned as supposedly masculine.

Despite all of this, life on a smallholding was not progressive for everyone. Particularly for women who were married, moving into a rural community and adopting a self-sufficient life meant negotiating expectations of male and female roles within their own families. It was not always something that people explicitly thought about beforehand, and even if they hadn't grown up with rural areas' stereotypically patrilineal, patriarchal mindset, it could still seep in unnoticed. This caused problems of its own. Corynne experienced some of the same reactions as Jessica and Lesley did, recalling how 'I mean I was there the weird lady, um, at College Fach, who used the cement mixer'. But for Corynne, the unspoken expectations around what women should be doing had even more impact than those about what they should not:

> We didn't have any money, and we had got to a stage with four children, two teenagers, [my husband] was a plumber, worked his socks off but it wasn't enough. So I went back to work. . . . But my husband was so used to my being at home, doing everything, bearing in mind we had two cows, three goats and forty sheep. Three of the four children had grievously expected me to carry on doing all this, plus shifts at the hospital, and it all just fell to pieces.

As we saw with the expectations for women in farming families, the workload of rural women was regarded as less significant than the more traditionally male tasks. As in most households in the 1960s and 1970s, women were left responsible for housework and childcare, but farming wives were also expected to do the smaller-scale agricultural work such as milking, egg collection, and perhaps some of the domestic gardening, as well as looking after the farm's accounts and records (Leckie 1996: 309–310). Side-line businesses, such as selling home products in a farm shop, also tended to fall to women (Woods 2011: 218). The crafts discussed in the previous section were usually their jobs as well, as domestic textile and food production are traditionally female occupations (one might recall how these roles fell to Freya and Gillian, while George took on the more stereotypically 'manly' DIY tasks). Guides to self-sufficiency tended to be written by men, often in firmly masculine language, and a number of basic 'feminine' tasks are mysteriously omitted from their directions; in *The Complete Book of Self-Sufficiency*, one can learn how to do something as obscure as building a methane digester but not how to do the laundry without running water (Munroe 1979; Seymour 1976). Of the identifiably gendered figures which appear in *The Complete Book of Self-Sufficiency*'s illustrations, there are more than three times as many men as there are women (83 vs. 22), and the jobs they do are also gendered. Men are shown completing a variety of hands-on tasks, such as haymaking, butchery, and building a wind turbine. Women are relegated to helping out men and illustrating the instructions for pottery and spinning. The illustrations in *Home Farm*, published a year later, seem to be populated entirely by men (Allaby and Tudge 1977). So, while it may seem on the surface as though countercultural migrants were bypassing the inequalities of rural life, the reality was that it was still very much a man's world.

The organic farming movement, which was developing in Wales at around the same time as self-sufficiency enthusiasts were buying up smallholdings, has been criticised for neglecting to address issues of gender in farming despite its otherwise progressive nature (Sumner and Llewelyn 2011: 116). It seems that the self-sufficient countercultural movement may have the same problem. Despite its initial promise and the potential it had to disregard traditional gender norms in the countryside, many women in self-sufficient lifestyles found themselves still restricted by the expectations placed upon them. We have already learnt that one of the reasons for the countercultural movement's development was as a reaction against growing consumerism. One factor for this spike in economic growth was the fact that more women were now working outside the home, and so more of the tasks they used to be doing (like clothes making, food preservation, etc.) were being outsourced to others (Perez 2019: 241). Caroline Criado Perez points out that 'productivity hadn't actually gone up. It had just shifted, from the invisibility of the feminised private sphere, to the sphere that counts: the male-dominated public sphere' (ibid). Judging by the stories shared here and the evidence from the way the guides to self-sufficiency were written, it seems that in abandoning mass production and consumerism, the countercultural migrants pushed this productivity firmly back into the (female) private sphere.

Looking at the everyday lives of the migrants who chose to try and become self-sufficient (or as close to it as possible) shows that their lives were simultaneously a rejection against convention – against consumerism, the separation of work and survival, the expectations of rural womanhood, etc. – and, in many ways, an extremely traditional way of life. With their emphasis on seasons and natural rhythms, practical skills, and 'make do and mend' attitudes, it can be hard not to see this as a return to an idealised image of the past. Of course, as the excerpts in this chapter show, the reality was far from easy, and as with any lifestyle, it was difficult for migrants to constantly hold on to all of what the counterculture supposedly stood for. Conventional nostalgia was entangled with visions of utopia, and traditional women's roles clashed with a rejection against the expectations of rural womanhood. This idea of an entanglement between counterculture and mainstream culture is something that we will continue to explore in the next chapter, where we move on from people's day-to-day home lives and look at the kind of work they were undertaking elsewhere.

References

Agg, Jenny and Phillips, Martin (1998) 'Neglected Gender Dimensions of Rural Social Restructuring'. In Boyle, Paul and Halfacree, Keith (eds.) *Migration Into Rural Areas: Theories and Issues*. Chichester: John Wiley & Sons.

Allaby, Michael and Tudge, Colin (1977) *Home Farm: Complete Food Self-Sufficiency*. London: Macmillan.

Brandth, Berit (2002) 'Gender Identity in European Family Farming: A Literature Review'. *Sociologia Ruralis*, 42(3): 181–200.

Chen, Biye, Lewis, Michael J. and Grandison, Alistair S. (2014) 'Effect of Seasonal Variation on the Composition and Properties of Raw Milk Destined for Processing in the UK'. *Food Chemistry*, 158: 216–223.

Cherrington, John (1983) *A Farming Year*. London: Hodder and Stoughton.

Duckham, A. N. (1963) *The Farming Year*. London: Chatto & Windus.

Evans, David, Campbell, Hugh and Murcott, Anne (2013) 'A Brief Pre-History of Food Waste and the Social Sciences'. *The Sociological Review*, 60(S2): 5–26.

Ewing, Elizabeth and Mackrell, Alice (1992) *History of Twentieth Century Fashion*. London: B.T. Batsford.

Galani-Moutafi, Vasiliki (2013) 'Rural Space (Re)produced – Practices, Performances and Visions: A Case Study from an Aegean Island'. *Journal of Rural Studies*, 32: 103–113.

Gardiner, Michael E. (2006) 'Everyday Knowledge'. *Theory, Culture & Society*, 23(2–3): 205–207.

Gibbs, D. C. (1987) 'Technology and the Clothing Industry'. *Area*, 19(4): 313–320.

Hellmann, Kai-Uwe and Luedicke, Marius K. (2018) 'The Throwaway Society: A Look in the Back Mirror'. *Journal of Consumer Policy*, 41: 83–87.

Hinds, Joe and Sparks, Paul (2008) 'Engaging with the Natural Environment: The Role of Affective Connection and Identity'. *Journal of Environmental Psychology*, 28: 109–120.

Hughes, Annie (1997) 'Women and Rurality: Gendered Experiences of "Community" in Village Life'. In Milbourne, Paul (ed.) *Revealing Rural 'Others': Representation, Power and Identity in the British Countryside*. London: Pinter. Pp. 167–188.

Jones, E. L. (1964) *Seasons and Prices: The Role of the Weather in English Agricultural History*. London: George Allen & Unwin.

Kalekin-Fishman, Devorah (2013) 'Sociology of Everyday Life'. *Current Sociology Review*, 61(5–6): 714–732.

King-Smith, Dick (1988) *Sophie's Snail*. London: Walker Books.

Langlands, Alexander (2017) *Cræft: How Traditional Crafts Are About More Than Just Making*. London: Faber and Faber.

Leckie, Gloria J. (1996) '"They Never Trusted Me to Drive": Farm Girls and the Gender Relations of Agricultural Information Transfer'. *Gender, Place and Culture*, 3(3): 309–325.

Little, Jo (2002) 'Rural Geography: Rural Gender Identity and the Performance of Masculinity and Femininity in the Countryside'. *Progress in Human Geography*, 26(5): 665–670.

Little, Jo and Austin, Patricia (1996) 'Women and the Rural Idyll'. *Journal of Rural Studies*, 12(2): 101–111.

Majima, Shinobu (2007) 'Fashion and Frequency of Purchase: Womenswear Consumption in Britain, 1961–2001'. *Journal of Fashion Marketing and Management*, 12(4): 502–517.

Mason, Rachel (2005) 'The Meaning and Value of Home-Based Craft'. *The International Journal of Art & Design Education*, 24(3): 261–268.

Morris, Carol and Evans, Nick (2001) '"Cheese Makers Are Always Women": Gendered Representations of Farm Life in the Agricultural Press'. *Gender, Place and Culture*, 8(4): 375–390.

Munroe, C. J. (1979) *The Smallholder's Guide*. London: David & Charles.

Peach, Andrea (2013) 'What Goes Around Comes Around? Craft Revival, the 1970s and Today'. *Craft Research*, 4(2): 161–179.

Perez, Caroline Criado (2019) *Invisible Women: Exposing Data Bias in a World Designed for Men*. London: Chatto & Windus.

Price, Linda (2010) '"Doing It With Men": Feminist Research Practice and Patriarchal Inheritance Practices in Welsh Family Farming'. *Gender, Place & Culture*, 17(1): 81–97.

Pyle, Robert Michael (2003) 'Nature Matrix: Reconnecting People and Nature'. *Oryx*, 37(2): 206–214.

Ridgeway, Cecilia L. (2009) 'Framed Before We Know It: How Gender Shapes Social Relations'. *Gender & Society*, 23(2): 145–160.

Riley, Mark (2009) 'Bringing the "Invisible Farmer" Into Sharper Focus: Gender Relations and Agricultural Practices in the Peak District (UK)'. *Gender, Place and Culture*, 16(6): 665–682.

Sandywell, Barry (2004) 'The Myth of Everyday Life: Toward a Heterology of the Ordinary'. *Cultural Studies*, 18(2/3): 160–180.

Saugeres, Lise (2002) 'The Cultural Representation of the Farming Landscape: Masculinity, Power and Nature'. *Journal of Rural Studies*, 18: 373–384.

Sennett, Richard (2008) *The Craftsman*. London: Penguin.

Seymour, John (1976) *The Complete Book of Self-Sufficiency*. London: Faber and Faber.

Seymour, John and Seymour, Sally (1973) *Self Sufficiency: The Science and Art of Producing and Preserving Your Own Food*. London: Faber and Faber.

Shortall, Sally (1992) 'Power Analysis and Farm Wives: An Empirical Study of the Power Relationships Affecting Women on Irish Farms'. *Sociologia Ruralis*, 32(4): 431–451.

Shortall, Sally (2014) 'Farming, Identity and Well-Being: Managing Changing Gender Roles Within Western European Farm Families'. *Anthropological Notebooks*, 20(3): 67–81.

Spence, Ian (2009) *Royal Horticultural Society Growing Through The Year*. London: Dorling Kindersley.

Sumner, Jennifer and Llewelyn, Sophie (2011) 'Organic Solutions? Gender and Organic Farming in the Age of Industrial Agriculture'. *Capitalism Nature Socialism*, 22(1): 100–118.

Tilley, Christopher (2006) 'The Sensory Dimension of Gardening'. *The Senses and Society*, 1(3): 311–330.
Turner, Will R., Nakamura, Toshihiko and Dinetti, Marco (2004) 'Global Urbanization and the Separation of Humans from Nature'. *BioScience*, 54(6): 585–590.
West, Elizabeth (1977) *Hovel in the Hills: An Account of 'The Simple Life'*. London: Faber and Faber.
Winston, Mark L. (1997) *Nature Wars: People vs. Pests*. Cambridge, MA: Harvard University Press.
Woods, Michael (2011) *Rural*. Abingdon: Routledge.

5 Making money

Employment and entrepreneurship

In this chapter, we turn our attention to the various ways that countercultural migrants' in Wales chose to earn money. Self-sufficiency and escape from external employment may have been ideal, but most people could not afford to abandon paid work entirely. They needed to find jobs in Wales to support themselves and their families, regardless of their lifestyle choices. The division between 'countercultural' and 'mainstream' becomes particularly blurred here; people may have been keen on countercultural ideals but unable or unwilling to live them all out in practice. Even commune inhabitants, generally seen as the most dedicated to countercultural principles, were frequently known to take on work outside the commune in order to support themselves (Halfacree 2006: 317–318). Here, I present the migrants' experiences alongside an introduction to the state of the employment market in Britain during the 1960s and 1970s. I am to outline the broad trends and changes of the era and begin to position this group of migrants within them.

In this chapter, I make much greater use of statistics than I do anywhere else, and this use of statistics comes with a few caveats. It is difficult to draw up an accurate list of exactly what jobs each contributor was doing, as the majority held a variety of different jobs at different times, with a great deal of fluctuation and overlap. The nature of the migrants' interviews means I do not have the same amount of employment data for each person; some spoke extensively about their jobs, others very little. Additionally, as I have only included jobs for the people I interviewed personally, the stories connected to their partners and families are absent. The statistics included are based only on the limited information I have been able to draw out of the narratives and should be treated as rough estimates only. Despite these drawbacks, it will help to put the following two sections into context if we can gain some understanding of how our group of migrants fitted into the bigger picture. Blending statistics with the stories that will come later is the best way to do this (Onwuegbuzie and Leech 2005: 384).

5.1 The rise of service industries

When Ian, whom we met in Chapter 1, finished his degree at Aberystwyth University, the next step was to find a job:

> For a while I worked on farms. And then that was, sort of early 70s and it was actually fairly easy to get jobs then. So, it's very different now, but I worked

DOI: 10.4324/9781003358671-8

on various farms, I was a petrol pump attendant, van driver, worked on building sites, all sorts of things.

His experience illustrates two important points: firstly, the state of the employment market in Britain during the 1960s and 1970s, and secondly, the variety of jobs countercultural migrants found themselves doing. During this era, Britain as a whole was indeed experiencing a period of high employment – the 'never had it so good' era that Macmillan is so well remembered for lauding (Sandbrook 2005). Employment levels had increased steadily through the 1950s and early 1960s (Owen et al. 1984: 469). Although they dropped slightly after this, unemployment remained low, hovering around 4–5% throughout the 1970s (Leaker 2009: 37).

This picture of stability does not tell the whole story, however, as it hides the tumultuous changes that were happening within the country. Wales was traditionally dominated by industrial and agricultural jobs, but changes in British industry meant that its economy declined throughout the post-war period (coming to a head with the miners' strikes of the 1970s and 1980s) (Cameron et al. 2002: 2). Welsh industry had been undergoing a process of diversification, with newer industries such as chemical production and processing, mechanical and electrical engineering, and vehicle production becoming increasingly present (Morris 1987: 209). But even with diversification, the decline couldn't be halted entirely. Between 1963 and 1983, the proportion of British jobs located in industry dropped from 46% to 34% (Hall 1987: 97). With agriculture also rapidly declining, the situation in Wales was far worse than the rest of the country (Johnes 2012: 150). The industrialised south of Wales experienced a brief period of revitalisation in the 1960s but failed to maintain this through the 1970s. Even in the 1960s, it was still regarded as a problematic area in terms of growth (Lovering 1985: 86). For migrants moving to Wales, there was a whole new employment landscape which they had to learn to negotiate if they were to survive.

Only two of the contributors worked in factories themselves, though many more had close family members employed in industry. It was discussed in Chapter 2 how various development initiatives were set up in rural Wales in order to encourage growth in the area, with 83 new factories opening between 1957 and 1973 (Halfacree and Boyle 1992: 11–14). This served its purpose of attracting people to the area and meant that even countercultural migrants often ended up in factory jobs. Heather, whom we met earlier, was one of those attracted by the new employment schemes. Her husband spent six months working in a carpet factory in order for them to claim government money for relocation to Wales:

[W]e relocated to Llandrindod Wells, which is a um, bit of the Cambrian inland. And um, he worked in a carpet factory, which was horrible, and then ditched it as soon as the six-months was up and took other work.

For the most part, though, the migrants tended towards other areas of employment. Figure 5.1 shows a breakdown of the types of jobs the contributors reported doing, categorised according to the Office for National Statistics' Standard Occupation

Classifications (SOC) (Office for National Statistics 2010). They are dominated by professional occupations, followed by skilled trades, associate professional and technical occupations, and managers, directors, and senior officials. These top four categories can be further broken down into a series of SOC subcategories (Figure 5.2).

This reveals a number of interesting trends. The dominance of textiles, printing, and other skilled trades is a reflection of the migrants' enthusiasm for crafts and practical skills. There are a large number of managers and proprietors too, and this is something we'll look at more closely later on. Despite tensions over the Welsh language (discussed more fully in Part 3), the biggest single employment category for the countercultural migrants was teaching and education. Indeed, many of the categories comprise service occupations of some kind, not just education but also health care and various public services.

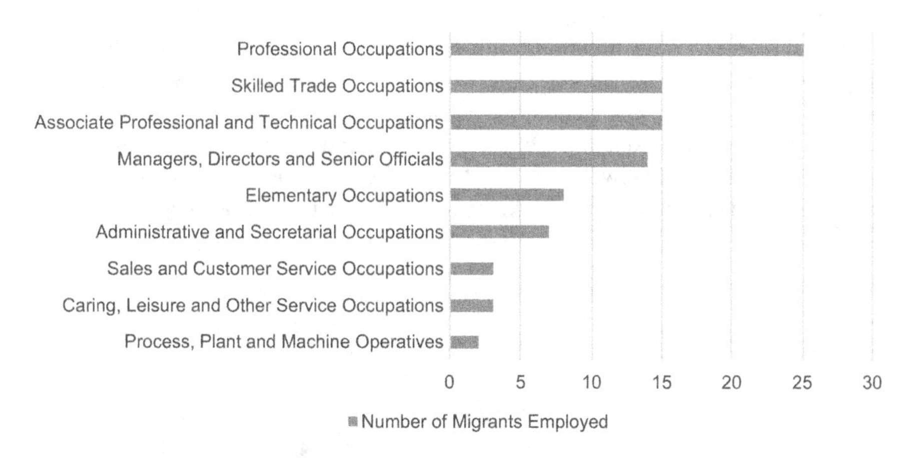

Figure 5.1 Total breakdown of interviewee occupations

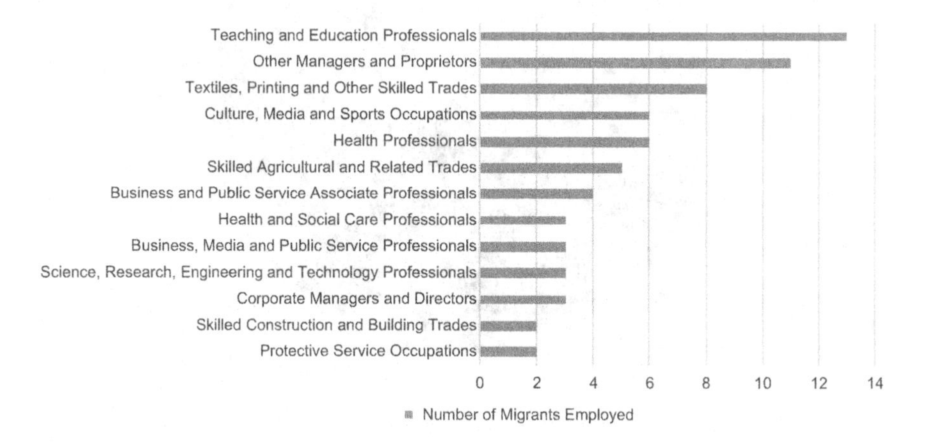

Figure 5.2 Breakdown of top interviewee employment sub-categories

These service jobs were plentiful since consumer services were something Wales was badly in need of; the absence of such facilities was still being remarked upon in research as late as 1985 (Lovering 1985: 99). Between 1971 and 1981, there was a surplus of jobs available in rural mid and West Wales (Owen et al. 1984: 474). This was partly because manufacturing jobs were faltering more slowly than elsewhere, but mainly because of the rapid growth in service industries. They became the dominant employers in Wales in 1971–73 (Williams 1982: 539). By 1981, 61% of all UK employment was in the service sector (Howell 2007: 92). The shift towards service industry work was connected to a simultaneous shift towards part-time work. This was increasing dramatically in Wales at this time and was particularly strong in the education and health sectors. Part-time employment in education increased by 55% in Wales during the 1970s, and part-time employment in health services increased by 71% across the UK during the same period (Townsend 1985: 323 & 327).

The shift away from traditional full-time industrial employment had implications for women's ability to enter the workplace. Employment in Wales had historically been male-dominated. As it became an overwhelmingly industrialised nation, women were excluded from the newly created industrial employment in mines, quarries, and steelworks. In 1961, only 28% of women in Wales were officially recorded as in or seeking work, the lowest rate across Britain (Williams 1982:

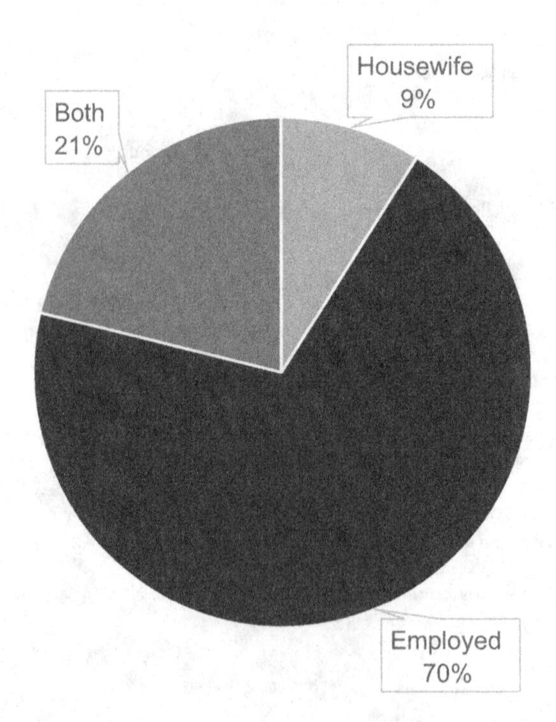

Figure 5.3 Breakdown of female interviewee employment

530–531). By 1975, however, this number had gone up to 38% and would only continue to increase (Cameron et al. 2002: 2). Of the women who contributed to this project and who gave some information about their employment history, almost 70% reported being more-or-less constantly in employment, with a further 21% reporting returning to work once their children were old enough (see Figure 5.3).

The countercultural migrants' jobs reflect overarching trends of the time: the decline of industry, growth of service industries, and increase in female employment. They also reflect the interest countercultural migrants had in crafts and practical skills. They may have frequently chosen jobs out of practical considerations more than anything else, but their choices also meant that they were providing necessary services for their local communities. This is something we will be taking a closer look at in the next section, where we focus on the migrants who were taking on managerial and proprietorial roles.

5.2 Starting a business

As shown earlier, many of the migrants to Wales went on to start their own businesses. This includes not just those who are listed as proprietors in Figures 5.1 and 5.2 but also many from other categories – craftspeople and tradespeople in particular tend towards self-employment. It should come as no surprise to learn that entrepreneurship was encouraged by John Seymour. His influence has already been well established, and this was just another area where people were inspired to follow his advice, even if it was not always the most practical. An excerpt from his foreword to Elizabeth West's *Hovel in the Hills* shows the sheer range of jobs Seymour had encountered:

> Bee-keeping; designing and selling a range of picture post-cards; publishing a series of small guides to the countryside; making wooden soup bowls on a lathe; building boats; making violins, spinning wheels, jewellery, leather objects; potting; weaving; engraving on slate; doubling as a landscape artist and a fisherman; illustrating botanical treatises; running a craft shop; running a second-hand tool shop; dentistry; doing fencing on contract; painting shop and van signs; jobbing building; nursery gardening; market gardening; sheep shearing: these are just a few of the solutions to this problem of earning a living *where you want to live* among people I know closely in my immediate vicinity that spring to mind without thinking about it very much.
>
> (Seymour 1977: 8, emphasis in original)

In reality, self-employment was not always as simple as Seymour made it sound. In the post-war period, the national focus was on public-sector initiatives rather than private-sector entrepreneurship. Money was invested in nationalising industries and protecting jobs at the expense of new businesses, and Britain's rate of new company foundation was the lowest in the world (Godley and Casson 2010: 258). Though there were some changes during Edward Heath's government in the early 1970s, the British population maintained an extremely low opinion of entrepreneurship, widely viewing it as a way for the unemployed to avoid getting 'real' work (Oliver and Pemberton 2004: 428; Godley and Casson 2010: 265). As with so many other things, the countercultural migrants' tendency for entrepreneurship and self-employment initially put them in the minority.

In the 1980s, this changed dramatically. Margaret Thatcher's government abandoned the public-sector focus of its predecessor, and by 1987, they had passed over 100 pieces of legislation promoting the creation and success of small businesses (Anderson et al. 2000: 11). Self-employment levels increased, rising from 9.7% in 1981 to 14.1% in 1990 (Robinson and Valeny 2005: 432). The experiences of the countercultural migrants cut across the stories of both decades, influenced by both sets of attitudes.

To understand more about the role of entrepreneurialism in these migrants' lives, we turn to the experiences of another contributor, Annie. She had spent time travelling around the US, Mexico, and Canada with her husband during the 1970s, and during this time Annie learnt leatherworking skills. On their travels, they earned money by working in orchards and selling various handmade products: bags, sandals, boxes, and so on. But on returning to Britain, they realised that, having sold most of their belongings in order to afford the trip, they were now without any money or place to live, though they did have a useful selection of new crafting skills. Two friends living in Wales offered to let the couple live in their barn as an inexpensive solution, and this was how they came to arrive in the area.

> We set up in the barn, there was a candle-making workshop, a little, little leather-making workshop and we started, started making things and it proved very easy to sell them. And uh, we went to a little trade fair, just to see, you know, if there was a market for what we were doing. And yes, loads of little shops in Wales seemed to want to buy what we were doing. So we thought, 'ok, let's go for it'. But obviously it wasn't going to work living in somebody's barn, so we started looking for a house, and at this point I think we made a conscious decision to stay here.

Money made from selling their products was initially supplemented by selling vegetables and Christmas trees grown on the land they purchased. This sort of supplementary business was common among countercultural migrants, many of whom made additional income this way; it tied in well with their desire to separate themselves from mass consumption and production. Supplementary work in building and house repairs was also popular. Again, this was a way of subverting mainstream practices. Linda Clark observes that self-employed builders were able to work 'in a manner not dissimilar to the artisan of the eighteenth century' (2005: 484).

The craft business was always at the core of Annie's life in Wales. In the late 1980s, her husband left her, and the business shifted away from leatherwork as she became interested in jewellery production:

> I suddenly thought 'oh, wouldn't it be nice if I could make jewelleries with Celtic knots on them'. And so I started a quest for beads with, beads with Celtic knots on them, and I couldn't find any. Couldn't find any anywhere, and it's kind of a bit early for the internet. Um, but you know there was one out there

> and you could look a bit. And I got catalogues, and I phoned people up, and I, you know wherever I went I looked in bead shops, but there really wasn't any such thing so I thought 'oh, ok. Perhaps I'll have a go at making them'.

According to Bögenhold and Staber, there are two main reasons for choosing self-employment: a desire for autonomy or economic necessity. They argue that for the person in search of autonomy, 'the motivation to become self-employed is based less on economic necessity than the desire to leave a stifling or otherwise unsatisfactory work situation' (1991: 226). This definition resonates well with the image of countercultural migrants as escaping the constraints of mainstream society. Annie was very keen to avoid living the kind of life her parents had, and the ability to be self-employed was a key part of her escape from it:

> I certainly felt having been raised in suburbia, with suburban ideas and . . . that I didn't, I didn't want that. . . . So to get away from it all was good, 'cause we'd found ways that we could support ourselves and not be a burden on society or family or anybody else at all.

Self-employment is one of the best ways to achieve a high level of autonomy at work, which in turn increases the amount of control people feel they have over their lives (Ross and Mirowsky 1992: 228). For countercultural migrants like Annie, it was a natural choice.

The beading business became highly successful, with its products sold on an international scale. The Welsh influence was a strong part of Annie's branding:

> I was selling the story of, of Wales and my journey and you know my little, my little tiny experience. So when people in America bought into what I was doing, I was sort of trying to put Wales on the map.

The 1970s revival in crafting was closely linked, in many ways, to the promotion of regional identity (Peach 2007: 245). Annie's craft business was part of this; she found a way to make her living by capitalising on her skills and the story of her move. Businesses like this would become an important part of rural economies and a key element of tourism in Wales, and Annie's was one of the earliest (Dlaske 2015: 243).

Helen's story is another example of the way countercultural migrants could go on to form their own businesses, but her experience was very different from Annie's. Helen moved to Wales in 1976 with her husband and baby son to work as wardens at the Youth Hostel in Crickhowell. After two years, they moved on to a different Youth Hostel, this time in Dolgellau. But by 1983, the family was tired of life as wardens:

> 'cause you sort of, you've got all the disadvantages of being self-employed because you're um, you know, you're just there out, you live on your own. . . . Um, but you've got the disadvantage of it's not your own business so you can't actually do your own thing.

It was around this time that Helen's husband became interested in computers. The 1980s saw a boom in British home computing, with the number of home 'micro-computers' quadrupling in 1982 alone (Lean 2012: 547). The BBC Model B was being introduced to schools, and it was this that gave Helen's husband the spark of an idea. He was going into schools as part of their hostelling work at the time, getting kids interested in the outdoors:

> He was doing field trips and he wanted to, the kids to be collecting data and analysing it. And he thought 'if I had a computer' . . . we didn't have a computer it was too early then, but there was a programmable calculator thing which he'd rather taken to. So he saw the BBC advertise and thought 'well this is just the thing, I'll get one and I'll buy software, and we can go out, we can collect data, we can measure river flows and measure slopes, those kinds of things, put all the data in and analyse it'. But there were no programs that did this.

Helen's husband had identified a useful gap in the market, and so the two of them wasted no time in coming up with a solution:

> Like often happens you end up writing the thing that you wanted to buy, do your own. . . . And so we started, he was writing programs and teachers were saying 'oh wow this is marvellous', 'cause they'd come back in the evening and get the computer out and analyse it, 'oh where did you get this program from?'. So we were first of all giving them away, and then he started, we started selling them. And then got to a point in 1984 where we were fed up of being wardens, 'cause cooking for up to 60 people is no joke, it's stressful.

They described themselves as being on the edges of the counterculture at this time, not totally immersed but still heavily influenced, and were keen to continue living in rural Wales. Programming offered them a way out of wardening, but one which wouldn't require them to move away. This is an example of entrepreneurship driven by opportunity recognition and innovation. Annie also exemplifies this mindset. It is one of the most reliable ways of developing a business, and something countercultural migrants seem particularly drawn to (Devece et al. 2016: 5369). Again, there is a connection that can be drawn between this and the countercultural ethos. Business historian Andrew Godley describes how 'Innovation requires vision and commitment – vision to imagine an alternative world in which the innovation has taken place, and commitment to mobilise resources to realise the vision rather than to just sit back and fantasise about it' (2010: 366). In choosing to move to Wales in search of a better life, countercultural migrants had already been through this process of vision and commitment once; it's not so surprising that they might do so again. A lot of the research into entrepreneurship focuses on urban areas, but many of the countercultural migrants turned to starting their own businesses as a way of maintaining rural life (McElwee and Smith 2014: 308). If you did not want to work in the dominant rural industries, this was often one of the best alternatives.

Alan moved to Wales for very different reasons, having been appointed as a consultant at Bronglais Hospital in Aberystwyth. The entrepreneurship in this case came not from him but from his wife:

> She was working, she was a qualified accountant, and was working, had a very good job in Cardiff. And as is often the case, or was the case, the wives followed the husbands. And although, you know, she had quite an important job in um, Glamorgan, for South Glamorgan County Council as um, for finance, in the finance department. So when we came to Aberystwyth she basically was out of a job. But fortunately she got a job as a deputy director of finance in Ceredigion County Council. Which she tolerated for about two years and became rather frustrated, and decided that she would go into business. And she opened an antique shop in Terrace Road, because we had always had an interest in antiques.

Alan's wife's story, with its initial theme of following her husband and his career at the expense of her own, is a common one. Women living in Wales at this time were particularly unlikely to obtain high-level jobs (Boyle and Halfacree 1994: 47). For Alan's wife, starting a business was a successful way of integrating as well as making money, with Alan noting how: 'that then made her feel very much more settled in Aberystwyth, and of course she loves it now and wouldn't want to leave'.

Antiques shops were a surprisingly popular business choice among the people interviewed, with at least four of them going into the antiques trade at some point. Others set up similar buying and selling enterprises, importing goods (clothing, jewellery, artwork, etc.) from abroad, in particular from Afghanistan (thanks to its connection with the counterculture via the hippy trail). Self-employment and entrepreneurship are not necessarily the same thing, but most of our self-employed group show a great deal of entrepreneurialism (Bennett et al. 2020: 8). Starting their own businesses enabled the migrants to carry their countercultural ethos through their careers. It granted them a greater level of control over their lives, separating them from mainstream culture in yet another way. It was also a way of finding satisfying employment in Wales, both for those who were unwilling or unable to enter the dominant service professions and for women whose previous careers suffered from the move.

5.3 'Couldn't buy garlic': the significance of wholefoods

There is a final major group of businesses started by countercultural migrants to Wales which I have not yet discussed: wholefoods shops. Since this was such an integral part of both their lives and the wider counterculture, it will be the subject of this section. Wholefoods were a key part of the counterculture of the 1960s and 1970s, and some of the biggest stereotypes surrounding countercultural migrants are associated with diet. Many countercultural migrants to Wales ended up founding or working in wholefoods shops or cafés. Even those who did not work in these shops were influenced by their presence, either because they purchased their products out of interest in the wholefoods diet, enjoyed the increased range of food choices they provided, or valued their presence as community facilities. This section gives an overview of countercultural diet trends – vegetarianism, macrobiotics, and wholefoods in general – linking them to the wider food trends that were developing at the time. It shows why it was so important for countercultural migrants to start up these businesses themselves rather than relying on existing local facilities and explores their social side as well as their business side.

Wholefoods as a concept had been ambling around in the background of society for a long time; the first recorded usage of the term itself dates back to 1880 (Oxford English Dictionary 2020b). Organic agriculture emerged in Britain during the 1920s and 1930s (Goodman et al. 2011: 57). While there was a connection between the organic movement and the rise of wholefoods, Geoffrey Jones notes that there was initially little success in attempts to connect consumers with organic produce. In 1948, the Wholefood Society was founded by Frank Newman Turner, a farmer and journalist, but it failed to make any real impact. Eleven years later, in 1959, an organic shop called Wholefood was founded in London, but it failed to make a profit for many years (2017: 60). It was only in the 1960s and 1970s that wholefoods truly began to take hold. We came across this briefly in Chapter 1, when we learnt about Barbara and Josie's interest in macrobiotics and Lesley's history of vegetarianism. For them, the interest rose out of attempts to be healthier and more environmentally friendly and concerns over animal welfare. These kinds of feelings were common in the counterculture. Jo and her husband, whom we met in Chapter 3, were some of the many migrants who felt uncomfortable about the state of the mainstream food industry at that time. After arriving in Wales, Jo worked as a nurse, while her husband did part-time work as a farm labourer while studying agriculture. Both jobs brought them into conflict with the world of food:

> Because I'd been a nurse, all the time I was nursing I thought 'people are getting ill who don't need to get ill. They could live in a way such as they didn't end up in hospital'. That's where it came from for me. And my husband had been studying agriculture, and learning about how much calories you have to put in a pig to get how many calories out in meat, and he didn't like that way of looking at the natural world. So we were, we were coming at it from two different angles. But also from the political, the idea of food miles, and pollution, you name it.

None of these were new phenomena. Meat farming has never been a particularly efficient means of food production; some people have always suffered from poor diets, and by as early as 1939, Britain was reliant on imports for a huge amount of food: 50% of its meat, 70% of its cheese, and 90% of its cereals all came from overseas (Ritchie and Roser 2020; Collingham 2017: 237). During the 1960s and 1970s, however, these negative aspects of food production were rapidly becoming more central as the British food industry underwent a series of drastic changes (Probyn 2011: 104). Farms were now increasingly reliant on mechanisation and chemicals in order to meet the ever-increasing consumer demand (Spencer 1993:

324–323). Meat consumption shot up from an average of 846 g per person per week in 1950 to 1,121 g in 1970 (Oddy 2013: 241). These changes, combined with increasing awareness of environmental pressures, meant more and more people were rebelling by turning to wholefoods, vegetarianism, macrobiotics, and so forth.

The problem was that foods like this were not widely available in Britain, and even when they could be found, they tended to be concentrated in urban areas. In 1968, there were 34 vegetarian restaurants in Britain, 16 of which were located in London. By 1978, the total number had increased to 132, but 52 of them were still concentrated in the capital (Spencer 1993: 322). If countercultural migrants to rural Wales wanted to keep buying their food more sustainably, they were going to have to start selling it themselves. The desire to live sustainably combined with the entrepreneurial spirit of the counterculture led many migrants to open their own shops. Jo and her husband were among them: 'We couldn't really get what we wanted to buy in terms of wholefoods. We were part of the early wholefoods movement, and . . . we decided to set up a wholefood shop in Aberystwyth.'

One of the reasons why wholefoods were so difficult to obtain in Wales but more common in urban areas was because they were dependent on a variety of imported international foods; lentils, brown rice, miso, soy sauce, etc. were all staples of the countercultural diet (Lau 2015: 62). This was partly because of the impact of macrobiotics, which had strong Japanese influences, and partly because countercultural migrants had often developed a taste for international foods when travelling on the hippy trail through Afghanistan and India (The Macrobiotic Association; The Independent 2011). Josie and her husband opened their health shop in Lampeter after returning to the UK from Afghanistan, and these kinds of new ingredients featured heavily:

> We used to drive up to London, my husband used to drive up to London in our van to buy whole foods from community suppliers in Camden, um, Camden Town. . . . We started off with a very small selection of um, whole foods. You know, beans and honey, and the things that we liked to eat because we'd travelled. Like halva, like olives and garlic, we introduced garlic. And um, so yeah we have had an influence because we were probably one of the first, I mean there was a health food shop in Aberystwyth, called The Friendly Stores, I think we, we were one of the first in Wales, basically.

The menu of the first macrobiotic restaurant to open in London, Seeds, established in 1968 by American brothers Craig & Greg Sams, who are credited with introducing macrobiotics to the UK, illustrates the centrality of imported ingredients (Crace 2009). It makes heavy use of brown rice and tahini and also features what were then exotic dishes like falafel and umeboshi plums (Sams n.d.). In urban areas during the 1960s and 1970s, international foods experienced a sudden increase in availability and popularity. The arrival of migrants from the colonies and elsewhere led to the opening of thousands of diverse restaurants and grocers in the areas where they settled. For members of the urban counterculture, the 'eating of Indian and Chinese food almost became the culinary equivalent of the sexual revolution' (Panayi 2008: 37 & 217–218).

In rural Wales, though, there was very little international migration. The exception to this were the Italian migrants to the Valleys, who in 1963 formed the second-largest non-British migrant group in Wales (after the Irish). But their numbers had begun to decline by the 1970s, and most other migrant groups were concentrated in industrial towns (Giudici 2014: 1413–1414). The lack of immigration meant that the international restaurants and grocers which were proliferating elsewhere had not yet arrived in rural Wales, and the products they provided remained out of reach. Megan, a newcomer to our story, arrived in Wales with her young family in 1973. Soon after moving, she encountered problems at the local grocers:

> I said I wanted some garlic to make ratatouille, garlic, peppers, courgettes and aubergine. And he said 'I'm not going to order it unless you can guarantee that I'll sell so much.' . . . When I first came here often you couldn't buy a lemon in Llani let alone garlic.

If garlic was out of reach, then miso was not even worth thinking about. Likewise, Judith and Catherine (who we met in Chapter 1) also struggled with the food when they moved to Wales, with Judith recalling that:

> I did miss some of the things Cathy's referring to, which is access to foreign food. Or food that we were used to, the sort of, kind of vegetables that we were used to being able to buy. Things like peppers for example, just ordinary peppers.

Because none of these foods were easily available in Wales, migrants who started selling them had a better chance of attracting non-countercultural customers; you did not need to be a hippy to want to sample international foods.

Louis, another new character in our story, moved to Wales in the 1960s after purchasing a dairy farm in Carmarthenshire. He and his family supplemented their income by selling homegrown produce and found it more of a local novelty than expected: 'we grew crops that they had never heard of, started selling vegetables at the local market. And, uh, people would come up and say, "what's this?", a red pepper or something. They just knew about potatoes and carrots'. By selling these exciting new products, countercultural migrants to Wales were able to tap into the freedom and rebellion they represented while simultaneously introducing the country to a range of new tastes and flavours that would otherwise remain unavailable.

These shops were frequented by members of the counterculture and locals alike, enabling other migrants to enjoy the foods they'd had in the cities while simultaneously introducing these foods to Wales. But the shops were more than just commercial spaces. Small, locally owned businesses are an important part of community infrastructure, helping to provide residents with a sense of place. The wholefoods shops started by migrants in Wales were no exception. For Jo and her husband, this was a core part of their business:

> The shop was far more than a food shop, it was really, we wanted it to be a political statement and an information centre, an exchange and a place for

> meetings and it was all those things, and so we were very, we were quite politc—well we were very politically active, so the shop was our base for that. . . . We did organise marches, for example to Greenham . . . and before that to Torness in Scotland, where they were building a nuclear power station. So we felt very linked in to protest movements all over the place.

Likewise, when Josie and her husband started up their shop in Lampeter, the social side was crucial, helping to draw together the local countercultural community:

> My husband's a musician, he's a singer-songwriter, and so that was another strand to what we've done. So this place has always been like a hub for people to meet one another, and for like-minded people to join together and, you know, meet. We were, we had one of the first meetings about growing organic foods here.

Smith and Sparks argue that, in addition to supplying products and services, small local shops play the following roles:

- providing a sense of diversity and choice to the consumer
- developing economic links with other businesses
- providing employment
- bring a sense of dynamism to the area

(2000: 205 & 207)

Local shops act as hubs of the community, places for people to meet, chat, and seek advice in an informal setting (Clarke and Banga 2010: 191). But the wholefoods shops in Wales came at a time when small local shops were beginning to decline, replaced by new American-style supermarkets. In 1961, there were 572 supermarkets in Britain. By 1969, this number had increased to around 3,400 (Alexander et al. 2009: 526). Through starting up these local businesses, countercultural migrants were not just spreading new foods but also rebelling against the rising tide of mass consumption and consumerism by keeping shopping small and sociable.

As this demonstrates, the significance of the wholefoods shops migrants started in Wales extended beyond simply providing food. They were an embodiment of countercultural interests in food politics and health, giving these ideas a physical presence in Wales. They were a way of spreading these ideas further by providing advice and acting as meeting places, as well as selling ingredients. They helped to introduce new foods to Wales, including ingredients from Japan, Afghanistan, India, and elsewhere, thanks to the international influences on countercultural diets. They acted as social hubs, providing an alternative to new American-style supermarkets and helping to maintain a sense of community. So significant were they that many of these shops and cafés continue to operate today. Through them, the countercultural migrants had a lasting impact on the commercial landscape of rural Wales as well as on the area's eating habits.

The blending of mainstream and counterculture that we began to observe in the daily lives of self-sufficient migrants becomes even more apparent when looking at the kind of work the countercultural migrants to Wales were doing outside the home. At the same time, many of the jobs they undertook shared a range of distinctive features which reflected the countercultural ethos. This group was adept at incorporating their lifestyle choices into their employment, seeking out work which focused on practical skills, spreading countercultural ideas and practices through food, and providing services to benefit their communities and bring them closer to the utopian ideal. They may not have been conscious of these decisions at the time; most were probably just following their own individual interests with little thought for the bigger picture, but cumulatively, we see that although plenty of mainstream ideas were apparent in their work lives, there was always a strong undercurrent of alternative beliefs.

References

Alexander, Andrew, Nell, Dawn, Bailey, Adrian R. and Shaw, Gareth (2009) 'The Co-Creation of a Retail Innovation: Shoppers and the Early Supermarket in Britain'. *Enterprise & Society*, 10(3): 529–558.

Anderson, Alistair R., Drakopoulou-Dodd, Sarah and Scott, Michael G. (2000) 'Religion as an Environmental Influence on Enterprise Culture: The Case of Britain in the 1980s'. *International Journal of Entrepreneurial Behaviour & Research*, 6(1): 5–20.

Bennett, Robert J., Smith, Harry, van Liesout, Carry, Montebrune, Piero and Newton, Gill (2020) *The Age of Entrepreneurship: Business Proprietors, Self-Employment and Corporations since 1851*. Abingdon: Routledge.

Bögenhold, Dieter and Staber, Udo (1991) 'The Decline and Rise of Self-Employment'. *Work, Employment & Society*, 5(2): 223–239.

Boyle, Paul J. and Halfacree, Keith H. (1994) 'Service Class Migration in England and Wales, 1980–1981: Identifying Gender-Specific Mobility Patterns'. *Regional Studies*, 29(1): 43–57.

Cameron, Gavin, Muellbauer, John and Snicker, Jonathan (2002) 'A Study in Structural Change: Relative Earnings in Wales since the 1970s'. *Regional Studies*, 36(1): 1–11.

Clark, Linda (2005) 'From Craft to Qualified Building Labour in Britain: A Comparative Approach'. *Labor History*, 46(4): 473–493.

Clarke, Ian and Banga, Sunil (2010) 'The Economic and Social Role of Small Stores: A Review of UK Evidence'. *The International Review of Retail, Distribution and Consumer Research*, 20(2): 187–215.

Collingham, Lizzie (2017) *The Hungry Empire: How Britain's Quest for Food Shaped the Modern World*. London: Vintage.

Crace, John (2009) 'The Wholefood Revolutionary'. *The Guardian*. Available at: www.theguardian.com/lifeandstyle/2009/jun/03/gregory-sams-wholefoodvegetarian-interview. Accessed 15 January 2020.

Devece, Carlos, Peris-Ortiz, Marta and Rueda-Armengot, Carlos (2016) 'Entrepreneurship During Economic Crisis: Success Factors and Paths to Failure'. *Journal of Business Research*, 69: 5366–5370.

Dlaske, Kati (2015) 'Discourse Matters: Localness as a Source of Authenticity in Craft Businesses in Peripheral Minority Language Sites'. *Critical Approaches to Discourse Analysis Across Disciplines*, 7(2): 243–262.

Giudici, Marco (2014) 'Immigrant Narratives and Nation-Building in a Stateless Nation: The Case of Italians in Post-Devolution Wales'. *Ethnic and Racial Studies*, 37(8): 1409–1426.

Godley, Andrew (2010) 'Culture, Opportunity and Entrepreneurship in Economic History: The Case of Britain in the Twentieth Century'. In Landström, Hans and Lohrke, Franz (eds.) *Historical Foundations of Entrepreneurship Research*. Cheltenham; Northampton, MA: Edward Elgar. Pp. 363–382.

Godley, Andrew and Casson, Mark (2010) 'History of Entrepreneurship: Britain, 1900–2000'. In Landes, David S., Mokyr, Joel and Baumol, William J. (eds.) *The Invention of Enterprise: Entrepreneurship from Ancient Mesopotamia to Modern Times*. Princeton; Oxford: Princeton University Press. Pp. 243–272.

Goodman, David, DuPuis, E. Melanie and Goodman, Michael K. (2011) *Alternative Food Networks: Knowledge, Practice, and Politics*. Abingdon: Routledge.

Halfacree, Keith (2006) 'From Dropping Out to Leading On? British Counter-Cultural Back-to-the-Land in a Changing Rurality'. *Progress in Human Geography*, 30(3): 309–336.

Halfacree, Keith and Boyle, Paul (1992) 'Population Migration Within, Into and Out of Wales in the Late Twentieth Century: 1. A General Overview of the Literature'. Migration Unit Research Paper 1.

Hall, Peter (1987) 'The Anatomy of Job Creation: Nations, Regions and Cities in the 1960s and 1970s'. *Regional Studies*, 21(2): 95–106.

Howell, Chris (2007) *Trade Unions and the State: The Construction of Industrial Relations Institutions in Britain, 1890–2000*. Princeton, NJ: Princeton University Press.

The Independent (2011) 'The Lonely Planet Journey: The Hippie Trail'. Available at: www.independent.co.uk/travel/europe/the-lonely-planet-journey-the-hippie-trail-6257275.html. Accessed 18 March 2019.

Johnes, Martin (2012) *Wales since 1939*. Manchester: Manchester University Press.

Jones, Geoffrey (2017) *Profits and Sustainability: A History of Green Entrepreneurship*. Oxford: Oxford University Press.

Lau, Kimberly J. (2015) *New Age Capitalism: Making Money East of Eden*. Philadelphia: University of Pennsylvania Press.

Leaker, Debra (2009) 'Unemployment Trends since the 1970s'. *Economic & Labour Market Review*, 3(2): 37–41.

Lean, Thomas (2012) 'Mediating the Microcomputer: The Educational Character of the 1980s British Popular Computing Boom'. *Public Understanding of Science*, 22(5): 546–558.

Lovering, J. (1985) 'Regional Intervention, Defence Industries, and the Structuring of Space in Britain: The Case of Bristol and South Wales'. *Environment and Planning D*, 3: 85–107.

The Macrobiotic Association. Available at: https://macrobiotics.org.uk. Accessed 20 May 2019.

The Macrobiotic Association. 'History of Macrobiotics'. Available at: https://macrobiotics.org.uk/history-of-macrobiotics/. Accessed 17 January 2020.

McElwee, Gerard and Smith, Robert (2014) 'Researching Rural Enterprise'. In Fayolle, Alain (ed.) *Handbook of Research on Entrepreneurship: What We Know and What We Need to Know*. Cheltenham; Northampton, MA: Edward Elgar. Pp. 307–334.

Morris, J. L. (1987) 'Industrial Restructuring, Foreign Direct Investment, and Uneven Development: The Case of Wales'. *Environment and Planning A*, 19: 205–224.

Oddy, Derek J. (2013) 'From Roast Beef to Chicken Nuggets: How Technology Changed Meat Consumption in Britain in the Twentieth Century'. In Oddy, Derek J. and Drouard, Alain (eds.) *The Food Industries of Europe in the Nineteenth and Twentieth Centuries*. London: Routledge.

Office for National Statistics (2010) 'ONS Occupation Coding Tool'. Available at: https://onsdigital.github.io/dp-classification-tools/standard-occupationalclassification/ONS_SOC_occupation_coding_tool.html. Accessed January 2 2020.

Oliver, Michael J. and Pemberton, Hugh (2004) 'Learning and Change in 20th-Century British Economic Policy'. *Governance: An International Journal of Policy, Administration, and Institutions*, 17(3): 415–441.

Onwuegbuzie, Anthony J. and Leech, Nancy L. (2005) 'On Becoming a Pragmatic Researcher: The Importance of Combining Quantitative and Qualitative Research Methodologies'. *International Journal of Social Research Methodology*, 8(5): 375–387.

Owen, D. W., Gillespie A. E. and Coombes, M. G. (1984) '"Job Shortfalls" in British Local Labour Market Areas: A Classification of Labour Supply and Demand Trends, 1971–1981'. *Regional Studies*, 18(6): 469–488.

Oxford English Dictionary (2020b) 'Wholefoods'. Available at: www.oed.com/view/Entry/430383?redirectedFrom=wholefood#eid33557035. Accessed 3 December 2020.

Panayi, Panikos (2008) *Spicing Up Britain: The Multicultural History of British Food*. London: Reaktion Books.

Peach, Andrea (2007) 'Craft, Souvenirs and the Commodification of National Identity in 1970s' Scotland'. *Journal of Design History*, 20(3): 243–257.

Probyn, Elspeth (2011) 'Feeding the World: Towards a Messy Ethics of Eating'. In Lewis, Tania and Potter, Emily (eds.) *Ethical Consumption: A Critical Introduction*. London: Routledge. Pp. 103–116.

Ritchie, Hannah and Roser, Max (2020) 'Meat and Dairy Production'. *Our World in Data*. Available at: https://ourworldindata.org/meat-production. Accessed 15 January 2020.

Robinson, Vaughan and Valeny, Rina (2005) 'Ethnic Minorities, Employment, Self-Employment, and Social Mobility in Postwar Britain'. In Loury, Glenn C., Modood, Tariq and Teles, Steven M. (eds.) *Ethnicity, Social Mobility, and Public Policy: Comparing the USA and UK*. Cambridge: Cambridge University Press. Pp. 414–447.

Ross, Catherine E. and Mirowsky, John (1992) 'Households, Employment, and the Sense of Control'. *Social Psychology Quarterly*, 55(3): 217–235.

Sams, Gregory (n.d.) 'Natural Food Story: Growing From Seed'. *Gregory Sams*. Available at: www.gregorysams.com/chaosworks/natural-food-history/. Accessed 30 March 2023.

Sandbrook, Dominic (2005) *Never Had It So Good: A History of Britain from Suez to the Beatles*. London: Abacus.

Seymour, John (1977) 'Foreword'. In West, Elizabeth (ed.) *Hovel in the Hills: An Account of 'The Simple Life'*. London: Faber and Faber.

Smith, Andrew and Sparks, Leigh (2000) 'The Role and Function of the Independent Small Shop: The Situation in Scotland'. *The International Review of Retail, Distribution and Consumer Research*, 10(2): 205–226.

Spencer, Colin (1993) *The Heretic's Feast: A History of Vegetarianism*. London: Fourth Estate.

Townsend, Alan (1985) 'Spatial Aspects of the Growth of Part-Time Employment in Britain'. *Regional Studies*, 20(4): 313–330.

Williams, Gwyn A. (1982) 'Women Workers in Wales, 1968–82'. *Welsh History Review*, 11: 530–548.

6 Away from work

Hobbies and interests

Just as important as what migrants were doing to earn a living was what they were doing for leisure in their spare time. However, because the countercultural ethos rebelled against conventional systems of work/leisure, it can be difficult to differentiate between the two. Take, for example, a migrant who has purchased a smallholding and spends most of their time working to grow food for themselves, but needs to supplement this by going out to work at a 'proper' job as well. Does this mean that the time spent on the smallholding is 'leisure', since it is not going out to work? A conventional distinction between work and leisure would say yes, but common sense tells us no – the time spent on the smallholding is more akin to the time spent on housework, and anyone can tell you that this is not leisure. To try and tease out the distinctions between the two, we begin with a discussion of what might constitute 'leisure time' as a whole, before moving on to a more detailed look at two of the activities countercultural migrants engaged with organised social groups and reading. The section closes with a discussion of countercultural spirituality as, although not strictly a form of leisure activity, this was another aspect of the migrants' lives which tended to be separate from their work.

6.1 Leisure time

The idea of 'leisure', as we know it today, did not fully take shape until the early twentieth century because of the same increases in industrialisation and loss of the direct link between work and survival, which the counterculture rebelled against (Veal 2005: 26). The average working hours for a male manuel worker in Britain had reached a record low in the 1970s, meaning that people had more time to spend on their own personal interests and hobbies (ibid: 29). However, clear definitions of this new 'leisure society' are few and far between, making it a difficult area to study (Veal 2011: 216). Early British research into leisure, which began to appear in the 1960s and 1970s, linked the concept of leisure to existing research on work, rather than considering it linked to the concept of lifestyle (Veal 2001: 359–360). Therefore, the separation between 'work' and 'leisure' was integral to this area of study from the outset. This is something we will come back to, as it is central to the countercultural migrants' experiences of leisure and hobbies. This section aims to provide a brief overview of the kinds of activities this group of migrants

DOI: 10.4324/9781003358671-9

did for fun. It focuses on organised social groups and activities, but this is mainly because people were more likely to speak about them, not because casual interests are unimportant.

It has been argued that 'hobbies', as we understand them today, only came into being in the nineteenth century. However, it has long been difficult to distinguish between activities undertaken as work and those undertaken for leisure. This is particularly the case regarding craft activities, such as sewing and gardening (Maines 2009: 19–22). In Chapter 4, we discussed the significance of crafts to the countercultural lifestyle, and their centrality is one of the things that makes leisure such a complex subject. To clearly differentiate work and leisure, studies on hobbies and leisure activities often make a distinction between 'serious leisure' and 'casual leisure'. According to sociologist Robert Stebbins, 'serious leisure is the systematic pursuit of an amateur, hobbyist, or volunteer activity that participants find so substantial and interesting that in the typical case, they launch themselves on a career centred on acquiring and expressing its special skills, knowledge, and experience' (Stebbins 2005: 200–201). Historian Peter Borsay also argues that there are two main forms of leisure, drawing a division between leisure activities that contain an element of play (defined as 'a self-contained activity without obvious consequences or significance') and those that can be thought of as 'serious leisure' (such as 'reading, gardening, craft hobbies, and such like') (2006: 6–8).

Applying this method of viewing leisure as either serious or casual to the countercultural migrants in Wales raises several questions. To what extent are their lifestyle choices 'serious leisure' and to what extent are they actually work? Where does 'casual leisure' fit into the picture? One way of getting a clearer idea of this is to compare the countercultural migrants' leisure activities with those of the average, 'mainstream' Briton of the time. In 1977, an official Household Survey was run to find out what leisure activities people in Britain were participating in, aside from watching television. A selection of the results is shown in Table 6.1.

As we can see, they are all fairly ordinary, centred around quiet domestic pursuits. Likewise, young people in the 1970s enjoyed, according to a separate survey, football, films, discos, pop music, board games, and drinking. Again, these are all

Table 6.1 Household survey of non-television leisure pursuits, 1977 (Marwick 2003: 207)

Activity	*% of Women Participating*	*% of Men Participating*
Going out for a meal or a drink	57	71
Listening to records or tapes	60	64
Reading books	57	52
House repairs & DIY	22	51
Gardening	22	49
Needlework & knitting	51	2
Countryside activities (walks, climbing, etc.)	27	31

much more conventional pursuits than the stereotype of rebellious 1970s youth would lead us to believe (Sandbrook 2013: 22). The contributors to this project also reported liking these kinds of activities. Plenty enjoyed going to the pub, were part of local sports teams, and went out to listen to local musicians. In many cases, the thing which made the countercultural migrants' social lives distinctive was not the activities themselves but the fact that they often had little time to devote to them: the heavy workload which comes with self-sufficiency could make finding spare time difficult. Working hours may have dropped for the average Briton, but rejecting conventional life meant rejecting its benefits as well as downsides. Although as one interviewee noted, this was not necessarily a bad thing:

> as owners eventually of nearly 40 acres with river and boating activities, animals, bees and all the accoutrements of a small holding we were pretty busy at home and have tended to entertain friends here as we have space and a beautiful place.

For countercultural migrants, there was a greater blending of work and leisure than was generally the case at the time. In Britain, the separation between work and leisure was growing, but these migrants often found the two spheres becoming intertwined in unique ways.

When it came to organised forms of leisure, the situation was slightly different. In the post-war years, there was a shift in the way people socialised, as social groups became more commonly organised around people's interests and hobbies than before (Balderstone 2014: 150). Whereas in earlier times, people were more likely to socialise simply with those near to them, now, it was easier to find and meet up with others who shared similar interests. This gave countercultural migrants a way of engaging with others who shared their interests in alternative living and those who shared their general, non-alternative interests. For Maureen, whom we met briefly in Chapter 2, the University Women's Club was a key part of her early life in Wales:

> It was a women's club for, run by the British Council, for people from abroad. And as I'd just come from Canada they asked me would I join as well. And it was mainly people from Iran, Pakistan, India. So there was, and there was people from Australia and New Zealand. There was a real mixture of different cultures. . . . And it was wonderful, we had international dinners and people brought things from their own country. And it was mainly, it was for women who had come from abroad to study, and were quite often very isolated. And this was really nice for them to get together.

Participating in women's groups could be a valuable way of spending leisure time, especially as moving can be an isolating and stressful process for women in particular (Magdol 2002: 558). Caroline, another migrant who arrived in Wales in 1978, experienced a similar situation. Like Maureen, the main reason for Caroline's move was because her husband had recently acquired a job at the university –

in this case, as a librarian. She joined the drama branch of the College Women's Club (a different organisation to the one that Maureen was involved with):

> And um, pretty well straight away I was in one of their plays. Which was in Theatr y Werin, which was a bit of a culture shock, I hadn't acted in large spaces like this, and I probably didn't have the vocal capacity to do it very well, but I very much enjoyed it. And um, we did a play called *The Rape of the Belt*, which is about Heracles and Antiope and the Amazons and is very funny. And um, from then I was doing plays with them, and at that time there were two or three staff based drama groups of various ages, as I say the College Women's was a very long established one, there was also staff drama. And, I mean although it was called College Women's we just incorporated any men we needed into our plays.

These two examples hint at another aspect of the social lives of these migrants. The class they came from made a big difference in their level of engagement with social groups. People from middle-class backgrounds have been shown to be active in twice as many organisations, on average, as those from working-class backgrounds (Hall 1999: 438). Working-class migrants were more likely to rely on informal social networks, such as those discussed in Part 3.

Even organised socialisation was not immune from the blending of work and leisure that resulted from the countercultural lifestyle. From the 1970s onwards, traditional women's social groups, such as the Women's Institute, were in decline. On the other hand, the membership of environmental and other activist groups increased dramatically (Hall 1999: 421). Voluntary work, often connected to these causes, was also a popular way of spending spare time; in 1976, 17% of British people were regularly participating in voluntary work (ibid: 425). Caroline was one of them, taking an active role in a number of charities alongside her amateur dramatics.

> I was um, worked for the WWF [World Wildlife Fund], and um, for a long time I ran an annual sponsored walk, to raise money for them and, sort of going from Aber up to the cliff path and Clarach or beyond on the hills. . . . There was actually a representative here, or in this area, at that time, and so, that was part of their projects, you know, but I actually ran the walk, and organised sponsorships and all that. . . . So, so I worked with them for quite a long time. And various other charities, I'm a great collector on street corners.

We see the countercultural ethos seeping through to people's leisure time especially strongly when it comes to social groups associated with charity work, as through this social activities became a way of furthering a set of goals as well as spending time with other people. We've already discussed how Jo's shop was used as a base for political activism in Aberystwyth, and many other migrants were also active in various political organisations such as the Campaign for Nuclear Disarmament (CND). This organisation, in particular, brought large numbers of people together to publicise and protest against the risks posed by nuclear weapons (Burkett 2012: 627). Gay, whom we have not met before, was another one of them:

> We were quite active in things like CND, marching down to Greenham Common and yeah, bussing to Greenham Common, marching down to. . . . We did a march with some Buddhist monks down the coast road to, to an airbase. . . . We planted a cherry tree outside their gate, as a symbol of peace. Which they didn't appreciate. And I remember, I remember everybody sort of all lying in the road in Aberystwyth. I can't remember quite why, but that was something CND related.

When it comes to organised leisure, it seems that it's often the way in which countercultural migrants participate that reflects their lifestyle more than the activities themselves. They chose groups that reflected their interests, and the interests of any one individual are always more varied than the type of lifestyle we ascribe to them. Yes, they were sometimes involved in countercultural groups, but they were involved in other groups just as often. The time restrictions that came with trying to be self-sufficient or starting one's own business had the biggest impact. And while this is not a characteristic unique to the countercultural lifestyle, it does seem to be an important part of it.

6.2 Reading; not just the underground press

As far as 'casual leisure' is concerned, the easiest activities to separate from work are those that do not contribute to income or resource production. Reading and listening to music are good examples of this. These also happen to be areas where the counterculture has had a strong impact, through the underground press and blossoming music scene. Paying attention to the things migrants were reading and listening to can tell us a great deal about the counterculture and their position within it. It also illustrates the difficulty of trying to segregate what is 'mainstream' and what is 'counterculture'. Since the literature and music of the 1960s/1970s are both extremely large and complex topics, I have chosen to focus only on what migrants were reading, particularly in relation to the wider countercultural publication scene.

The most obvious branch of countercultural publications was the underground press. Sue, whom we met in Chapter 2, vividly remembers the impact of the underground press and related music scene, and its constant presence in her circle of friends:

> There was a whole generation of people . . ., who didn't accept that, you know, they were going to be channelled into this particular way of living, and that education, um, was just more brainwashing. You know it was more brain – the word brainwashing was used a lot, you know. And you had all the underground magazines coming out at that point. . . . You know they were always around. You know they were, people had piles of them everywhere, you know, and hundreds of records, you know, and sort of um, stereo tapes, and then quadraphonics.

In the 1960s, there was a dramatic increase in all forms of communication (Rycroft 2003: 104). The underground press was part of this. It documented British counterculture, being 'in effect the only viable institution created by the essentially anti-institutional counter-culture' (Nelson 1989: ix–x). The most famous of these papers tended to be quite London-centric, containing an eclectic mixture of alternative culture and anarchist-leaning politics. England's first underground paper, the *International Times*, was launched in 1966 (ibid: 45). It became known as *IT* shortly afterwards, following a legal dispute from *The Times* itself (Fountain 1988: 30). Other publications soon followed, including *Oz*, *INK*, *Friends* (later known as *Frendz*), *Gandalf's Garden*, *The Black Dwarf*, and *The Hustler* (Bratus 2012: 227). This type of magazine was known for its inflammatory, often bordering on obscene, content (Sutherland 1983: 3). A good example of the kind of material they contained was the cover of *IT*, Vol. 1, Issue 17, which proclaims that it contained articles on such topics as 'Cannabis Can Cure Acne', 'Consciousness is Addictive', and 'All Politics is Pigshit' (*IT* 1967).

The underground press rebelled against the categorisation of ideas that was prevalent in traditional media (Nelson 1989: 47 & 54). These publications were meant to be an eclectic combination of everything and anything. But, despite this, different magazines still catered to different audiences. Not many of the people I interviewed recorded reading the branch of magazines described earlier, as those publications tended to be quite urban-focused in reach and content, and of little practical interest to lifestyle migrants in Wales. Jo, on the other hand, remembers a different side to the underground press:

> I was reading a fair amount of . . . we subscribed to Peace News, which still exists, it's changed a lot, and in the 70s, it had quite an emphasis on lifestyle, on living simply, and growing things. So, I read that each time it came through the door, and we sold it in the shop. . . . And we, I read a magazine called Seeds [sic], might be not called Seeds but it might have been.

Peace News was established in 1936, with the aim of spreading news about worldwide peace and justice issues (*Peace News* 2018). Likewise, CND had its own newspaper, *Sanity*, which ran from 1961 to 1991 (British Library). These were very different from *IT*, *Oz*, and the other big names of the urban subcultures. Seed also fell into this category. Craig and Greg Sams, the same brothers who had introduced macrobiotics to Britain, worked with their father Ken to publish *Seed: The Journal of Organic Living* from 1972 to 1977 (Sams 2018a). In some ways, *Seed* contained similar content to John Seymour's books, and others in a similar vein. It contained regular articles from 'John Butler, Organic Farmer', with information about growing food without chemicals, along with suggestions for recipes and instructions on practical skills like food preservation and various types of crafts. It also went further, with many articles explaining the flaws behind conventional ways of doing things. Some examples include: 'Milk: the greatest cause of disease', 'Sugar: the £multi-million body wrecker' (*Seed* 1972a), and 'The Meat Protein Myth' (*Seed* 1972b). On the other hand, *Seed* also contained a lot of articles that came from a very different side of the counterculture – astrology, herbalism, and a column by 'Mr Ouija', the resident philosopher, all featured regularly (Sams 2018b).

One of the most influential countercultural writers of the era was E.F. Schumacher, whose classic revisioning of economics, *Small is Beautiful: economics as if people mattered*, was published in 1973. It sold over 700,000 copies, and the impact of his ideas is still being felt (Binns 2006: 218). Schumacher was opposed to the idea that economics can ever be a neutral, objective science, and argued against the use of fossil fuels and nuclear energy because of their finite nature on one hand and disastrous environmental impact on the other. He advocated for a Buddhist-inspired model of economics, based on minimising consumption and developing technologies as appropriate (Varma 2003: 115–117; Kirk 2001: 380). But, despite Schumacher's popularity and influence, a few interviewees reported reading his work. As with the urban underground press, the impact of its existence was felt only indirectly.

Lesley was a reader of this more political strand of the underground press:

> I subscribed to Spare Rib so once a month I'd get like a dose of feminism. Um, which is how I heard about the women's conference. And I did actually advertise in Spare Rib to see if there were any other women out there, but there weren't. And interestingly when my partner, my current life partner, moved to Wales she also advertised in Spare Rib, but I obviously didn't see that issue, otherwise we would have met a lot earlier. . . . But just little things like that just sort of keeping, you know, a bit of a thread, but, and Spare Rib was very London orientated at the time, so there wasn't a lot to relate to, but it was good to know that there was still feminist sort of consciousness going on out there.

Her story illustrates the contradiction inherent to the underground press: it was simultaneously a way of uniting the counterculture through access to a shared channel of communication and too strongly regional, lacking in any clear purpose or ideology (Bratus 2012: 244).

This is not to say that there was no point to the underground press at all. Magazines can help create a sense of community (Frith 2012: 227). Alternative magazines, catering to a very specific range of interests, are particularly good at this (Reader and Moist 2008: 824). In many ways, this is reflected in the experiences of migrants to Wales. Like many other women, Lesley was able to connect with likeminded people through *Spare Rib*, and it offered a way of finding out what might be going on (Smith and Quaid 2017: 6). Many other alternative publications offered similar opportunities, with events listings, letters pages, and classified ads all commonplace. People could read about the things that interested them, contribute where they wanted to, and maybe connect with other like-minded people in real life. They created a 'virtual community' – a way for a dispersed group of people to interact (ibid).

As positive as all of this sounds, the problems which plagued the underground press meant that it never fully lived up to its potential to unite the counterculture. The London-centric nature of most of the studies meant that, as Lesley recalled, it was difficult for migrants to Wales to relate to much of their content. Many underground publications were troubled by difficulties with finance and distribution; hence, they were less likely to make their way to Wales anyway – it was down to alternative shopkeepers like Jo to make sure they were stocked locally (Atton 1999: 53–54). Even when they were available, there was no unifying ideology to connect each different branch of publications; nothing explicitly linked *Spare Rib* to *IT*, or *IT* to *Seed*, and so on. Indeed, the lack of clear uniting ethos meant that it could sometimes be very difficult to differentiate between 'alternative' and 'mainstream' publications, as it was never clear exactly what they were an alternative to (ibid: 51).

By the time most of these migrants arrived in Wales during the mid-1970s, the golden age of the underground press was already over. Although many contributors did report reading some form of underground or alternative periodical, far more seem to have stuck to a combination of traditional newspapers and the classic books on alternative living that we have already come across, such as John Seymour's numerous works, Elizabeth West's *Hovel in the Hills*, Thomas Firbank's *I Bought a Mountain*, and so on. It is also important not to forget that just because people were part of the counterculture in some ways does not mean they were obliged to eschew mainstream culture entirely. By far, the most common things people reported enjoying reading were perfectly normal: popular science, murder mysteries, and adventure novels all featured regularly. The literature people were reading is one of the areas where the blending of cultures is most apparent – being part of the counterculture did not mean that you could only read countercultural publications, and even if you did, it did not necessarily mean that you were reading the same things as other members of the counterculture.

6.3 Quakers and Buddhists: alternative spiritualities

Spirituality is another area where the counterculture is particularly diverse. Ideas about alternative spirituality form a big part of the hippy stereotype. Traditional religion – which, in Britain, generally means Protestant (and usually Anglican) Christianity – did not fit into the hippy ideal. When one pictures countercultural religion, one thinks of astrology, fertility goddesses, and pagan festivals, usually involving Druids celebrating solstices at Stonehenge. The reality is in fact very different. There was indeed a tremendous range of diversity among the people I interviewed, but they covered the full spectrum of religions, and not just those considered 'alternative'. From atheists to agnostics and Anglican priests to Zen monks, countercultural religion encompassed far more than just astrology and paganism.

Religion in Britain underwent a seismic change over the course of 1960s and 1970s. Until this point, most of the British population had been Christians, most of whom were members of the Church of England. But, between 1960 and 1985, membership of the Church of England shrank by half (Davie 2015: 49–50). In his history of religion in twentieth century Britain, Callum Brown observed that:

> The sixties was the most important decade for the decline of religion in British history. Pop music, radical fashion and student revolt were witness to a

> sea-change in sexual attitudes and to the dismissal of conventional social authority. There was a cultural revolution amongst young people, women and people of colour that targeted the churches, the older generation and government. In this maelstrom, traditional religious conceptions of piety were to be suddenly shattered, ending centuries of consensus Christian culture in Britain. In its place, there came liberalisation, diversity and freedom of individual choice in moral behaviour. In every sphere of life, religion was in crisis (Brown 2006: 224).

This religious crisis did not automatically equate to a widespread renouncement of all religions though. Although the religious group which experienced the largest increase in numbers over the post-war years was undeniably those with no religious affiliation, these were not all die-hard atheists – many were agnostic or simply uncomfortable with labelling themselves (Woodhead 2016: 250). At the same time, there was also an increase in non-denominational Christians and those practising non-Christian religions as a whole (Davie 2015: 47).

It was in this rapidly shifting religious landscape that the counterculture of the 1960s and 1970s emerged and positioned itself. Although there was never a single, overarching countercultural form of spirituality, it was a constant presence. Magazines in the underground press regularly printed articles on various forms of spirituality as they related to the counterculture. *Gandalf's Garden*, which only ran for 6 issues in the late 1960s but was nevertheless influential, had a particular focus on spirituality with articles such as 'The Glastonbury Mystique: Jesus and the Druids' and 'Tibetan Monastic Centre' (Gandalf's Garden 2006). *Seed*, which was discussed earlier in this chapter, related spirituality to food and natural living, with features on topics like 'Supranature: Fusion of Science and the Occult' and 'Saturnalia: Feast of Fools' (Sams 2018b).

Countercultural spiritualism was a conglomeration of different religions and beliefs. Because of this they are difficult to define: New Age, Pagan, Mind-Body-Spirit, and Druid cultures all fall, in one way or another, into the category of 'alternative spirituality' (Harvey and Vincent 2012: 157). These were not new to Britain, but their popularity increased over the course of the latter twentieth century as they became (relatively) more socially acceptable. The Witchcraft Act 1736 was only repealed in 1951, illustrating the changing attitudes (ibid: 160). Among the contributors to this project, the most obvious influence on countercultural spirituality was Buddhism. In the 1950s, Buddhism had been something of an exotic novelty to most people in Britain. It wasn't unheard of – during the Victorian era, newspapers had been flooded with accounts of Buddhism as books and journal articles penned by travellers to the East abounded – but it certainly was not common (Franklin 2008: viii). From the 1960s onwards, however, there was a steady increase in the number of Buddhists in Britain. This was due to two main factors: Western converts to Buddhism, and the arrival of Buddhist immigrants and refugees from Asia (in particular those fleeing the Chinese invasion of Tibet in 1959) (Starkey and Tomalin 2016: 327). Many young countercultural enthusiasts had encountered Buddhism while travelling through Asia on the hippy trail or from the stories of those who had returned. It was seen as 'a religion providing a modern man with the answers to spiritual despair ensuing from rampant industrialisation, rationalism, secular-materialism, and consumerism' (Kay 2007: 6). This made it the perfect solution to the search for a countercultural spirituality.

We learnt in Chapter 1 that Lesley had converted to Buddhism very early on, long before moving to Wales, but for her, the link between Buddhism and the alternative lifestyle was clear and powerful:

> I suppose the self-sufficiency was, I mean you could sort of like link it in a way with, with politics and with Buddhism really because it's all trying to get to the fundamentals, of the sort of like the essence of things. Rather than all the superficial stuff that we sort of layer on top. And that's really important to me.

At the same time though, there were undeniably tensions between the self-sufficient lifestyle and Buddhist philosophies. Lesley was a vegetarian but still found herself occasionally needing to kill off the young male goats her herd produced (as they were not essential for milk production):

> I mean I know when I would have to put the kids down. I would always do a sort of ritual before my Buddha. Because I knew the, I knew the karma consequences of killing, but I also felt this was the most responsible thing I could do for the animal. So it was like, you know, I'd brought this creature into the world, I was fully responsible, and I was quite accepting of the consequences of ending its life.

This is a very different attitude to the matter-of-fact way that John Seymour's books were written, or the usual way-of-life mindset that one traditionally finds in rural areas.

Another common religious trend among the migrants was, perhaps surprisingly, the Society of Friends or Quakers. There are numerous similarities between Quaker and Buddhist philosophies, most notably not only in the values each place on silence and simple living but also in the emphasis on kindness (Bell and Collins 1998: 2). Klaus Huber, who conducted a study of Buddhist Quaker identity in 2002, observed how 'the mettã meditation, with its aim of developing loving kindness for all living beings, explicitly including one's enemies, brings to mind Jesus' words . . . and links with the pacifist tradition of Quakerism' (Huber 2002: 93).

Some people questioned the suitability of importing religions such as Buddhism. British faiths may have felt too close to home at first, but they became more attractive as time went on. Helen, introduced earlier, was strongly influenced by Buddhist ideas but was uncomfortable with adopting a faith completely alien to the land she was living in: 'There was very briefly a Buddhist group here but they did Tibetan, they were doing Tibetan chanting, and I'm not a Tibetan. . . . It's fine for a Tibetan monk, but I'm not a Tibetan'. As a solution, she eventually combined her Buddhist beliefs with those of the Quakers, their shared similarities lending themselves to this arrangement. I am trained as a Zen monk while staying in Thailand for a time but came to the same arrangement as Helen on his return to Wales:

> I also go to the Quakers, so, you know, although I am basically Buddhist I don't really like the label because I'm not sure that other people understand my . . . you know I read a lot of Christian literature and stuff and I feel it's really the same thing.

Blending multiple faiths was not an uncommon occurrence among the counterculture movement nor is it a phenomenon unique to it. In their account of alternative spiritualities in Britain, Graham Harvey and Giselle Vincent point out that

> the modern formation of alternative spiritualities in the West, especially since the late eighteenth and nineteenth centuries, meant that they were created by people who had a firm grounding in Christianity. Druids, and others, of the time were Christian and would not have seen that dual identity as incompatible.
>
> (2012: 165)

As popular as alternative spiritualities were, they were not obligatory for members of the counterculture. Many contributors identified as Christians, and for some, this formed an integral part of their identity. For Jim, who was introduced in Chapter 3, Christian faith was a constant presence and a way of integrating. Likewise, a large number of migrants identified as either atheists or agnostics (often shifting between various levels of the two). This could cause tensions in Wales. Jessica, whom we met earlier, had been brought up in the Church of England. As she grew older, she found herself drifting further away from the church, as many did, and eventually became 'a devout atheist'. But, in rural Wales, during the 1970s, atheism was all but unheard of: 'I heard people say, 'oh, she's a wicked atheist', as if being wicked and being an atheist were more or less the same, you know. And um, there were a lot of people there that were chapel, a few people were church, but um, not, not many atheists around'. Wales's religious landscape has always differed from England's, and so the religious upheaval of the 1960s and 1970s affected it in different ways and for different reasons. One study points out that:

> Wales is more characterised by religious decline and indifference, but chiefly because religio-cultural bonds – in this case bound up in the history of the Liberal Party and Methodism as carriers of Welsh identity – unravelled in the latter half of the twentieth century in parallel with traditional industries and gender roles.
>
> (Guest et al. 2012: 66)

Wales was used to religious uncertainty and division, the prominence of Chapels and Methodists in Welsh culture made certain of this, but the country was unused to outright rejection of religion.

Among members of the counterculture, the range of religious identities was incredibly diverse. Although the underground press may have publicised what seemed to be a particular form of alternative spirituality, when one starts unpicking this, it is revealed to be a rich and varied patchwork of customs, philosophies, and beliefs taken from an international range of influences. There was really no single, identifiable standard alternative religion. As the contributors to this project showed through their stories, individuals wove together their own beliefs from the ideas they were most drawn to, combining Eastern faiths, Celtic spiritualities, Christianity, and more. Because alternative spirituality was so diverse, there was no reason that one could not belong to the counterculture and remain a traditional Christian, as Jim did, or renounce religion entirely like Jessica. The choices people made may have affected how they chose to live out their lives, just like Lesley's awareness of the karmic repercussions of animal husbandry, but this was all part of the counterculture – a way for people to define their own identities on their own terms, regardless of what mainstream culture was telling them they should be.

Leisure and spirituality are two of the areas that the stereotypical 'hippy' image discussed in the introduction draws most influence from. Activist groups, the underground press, and alternative spiritualities all feature heavily in hippy stereotypes. But, as we have seen, the countercultural migrants' experience of leisure and spirituality was characterised by an extensive blending of counter and mainstream cultures. The contrast between the stereotype and the reality highlights the fact that the stereotype is simply that – a stereotype. It would seem that the migrants' stories suggest, yet again, that realistically all of their lives were a combination of both cultures, their interests disregarding the boundaries between them.

Looking at them from the outside, and with the benefit of hindsight, the everyday lives of countercultural migrants to Wales seem characterised by blurred contrasts: work vs. play, mainstream vs. alternative, and modern vs. traditional. Each individual story weaves together so many different strands that it's impossible to generalise and create an accurate depiction of the 'typical' migrant experience. All I can do here is share the stories, highlighting the recurring themes and variations within them and linking them to wider issues and events. However, doing so does show that there are some interesting commonalities. Tensions often arose between the desire to reject aspects of conventional life – such as the growing disconnection from natural rhythms, the valuing of money over time, or increasing industrialisation – and the need to find a way of surviving and thriving in a rural area. In their day-to-day lives, migrants blended together different elements of counter and mainstream cultures, sometimes following one closely and at other times pursuing different interests. As mentioned earlier, this is something the migrants were unlikely to be aware of at the time. People do not instinctively analyse their own lives, arbitrarily dividing them up and categorising them. It is only with retrospect that we can see these patterns become apparent.

References

Atton, Chris (1999) 'A Reassessment of the Alternative Press'. *Media, Culture & Society*, 21: 51–76.

Balderstone, Laura (2014) 'Semi-Detached Britain? Reviewing Suburban Engagement in Twentieth-Century Society'. *Urban History*, 41(1): 141–160.

Bell, Sandra and Collins, Peter (2014) 'Religious Silence: British Quakerism and British Buddhism Compared'. *Quaker Studies*, 3(1): 1–26.

Binns, Tony (2006) 'E. F. (Fritz) Schumacher'. In Simon, David (ed.) *Fifty Key Thinkers on Development*. London: Routledge. Pp. 218–223.

Borsay, Peter (2006) *A History of Leisure: The British Experience since 1500*. Basingstoke: Palgrave Macmillan.

Bratus, Alessandro (2012) 'Scene through the Press: Rock Music and Underground Papers in London, 1966–73'. *Twentieth-Century Music*, 8(2): 227–252.

British Library. 'Learning: Dreamers and Dissenters'. Available at: www.bl.uk/learning/histcitizen/21cc/counterculture/demonstration/sanity/sanity.html. Accessed: 23 January 2020.

Brown, Callum G. (2006) *Religion and Society in Twentieth Century Britain*. Harlow: Pearson Education Limited.

Burkett, Jodi (2012) 'The Campaign for Nuclear Disarmament and Changing Attitudes towards the Earth in the Nuclear Age'. *British Journal for the History of Science*, 45(4): 625–639.

Davie, Grace (2015) *Religion in Britain: A Persistent Paradox*. Chichester: Wiley Blackwell.

Firbank, Thomas (1940) *I Bought a Mountain*. Countryman Press.

Fountain, N. (1988) *Underground: The London Alternative Press, 1966–74*. London: Routledge.

Franklin, J. Jeffrey (2008) *The Lotus and the Lion: Buddhism and the British Empire*. Ithaca; London: Cornell University Press.

Frith, Cary Roberts (2012) 'Magazines and Community'. In Reader, Bill and Hatcher, John A. (eds.) *Foundations of Community Journalism*. London: SAGE. Pp. 223–236.

Gandalf's Garden (2006) 'Contents'. *Gandalf's Garden*. Available at: www.pardoes.info/roanddarroll/GGContents.html. Accessed 5 February 2020.

Guest, Mathew, Olson, Elizabeth and Wolffe, John (2012) 'Christianity: Loss of Monopoly'. In Woodhead, Linda and Catto, Rebecca (eds.) *Religion and Change in Modern Britain*. London: Routledge. Pp. 57–78.

Hall, Peter A. (1999) 'Social Capital in Britain'. *British Journal of Politics*, 29: 417–461.

Harvey, Graham and Vincent, Giselle (2012) 'Alternative Spiritualities: Marginal and Mainstream'. In Woodhead, Linda and Catto, Rebecca (eds.) *Religion and Change in Modern Britain*. London: Routledge. Pp. 156–172.

Huber, Klaus (2002) 'Questions of Identity among "Buddhist Quakers"'. *Quaker Studies*, 6(1): 80–105.

IT (1967) 'Cover Image'. *IT*, 1(17). Available at: www.internationaltimes.it/archive/index.php?year=1967&volume=IT-Volume-1&issue=17&item=IT_1967-07-28_B-IT-Volume-1_Iss-17_001. Accessed 25 January 2020.

Kay, David N. (2007) *Tibetan and Zen Buddhism in Britain: Transplantation, Development and Adaptation*. London: Routledge.

Kirk, Andrew (2001) 'Appropriating Technology: The Whole Earth Catalog and Counterculture Environmental Politics'. *Environmental History*, 6(3): 374–394.

Magdol, Lynn (2002) 'Is Moving Gendered? The Effects of Residential Mobility on the Psychological Well-Being of Men and Women'. *Sex Roles*, 47(11/12): 553–560.

Maines, Rachel P. (2009) *Hedonizing Technologies: Paths to Pleasure in Hobbies and Leisure*. Baltimore: John Hopkins University Press.

Nelson, Elizabeth (1989) *The British Counter-Culture, 1966–73: A Study of the Underground Press*. Basingstoke: Macmillan.

Peace News (2018) 'About Peace News'. *Peace News: For Nonviolent Revolution.* Available at: https://peacenews.info/about-peace-news. Accessed 21 January 2020.

Reader, Bill and Moist, Kevin (2008) 'Letters as Indicators of Community Values: Two Case Studies of Alternative Magazines'. *Journalism & Mass Communication Quarterly*, 85(4): 823–840.

Rycroft, Simon (2003). 'Mapping Underground London: The Cultural Politics of Nature, Technology and Humanity'. *Cultural Geographies*, 10(1): 84–111.

Sams, Craig (2018a) 'About Me'. Craig Sams. Available at: www.craigsams.com/about-me. Accessed 21 January 2020.

Sams, Craig (2018b) 'Seed Magazine'. Craig Sams. Available at: https://craigsamswp.wordpress.com/writings/seed-magazine/. Accessed 21 January 2020.

Sandbrook, Dominic (2013) *Seasons in the Sun: The Battle for Britain, 1974–1979.* London: Penguin Books.

Seed (1972a) 'Sugar: The £Multi-Million Body Wrecker'. *Seed* 1(5): 10–11. Available at: https://craigsamswp.files.wordpress.com/2015/03/seed-v1-n5-sept1972.pdf. Accessed 25 January 2020.

Seed (1972b) 'The Meat Protein Myth'. *Seed,* 1(3): 1–3. Available at: https://craigsamswp.files.wordpress.com/2015/03/seed-v1-n3-april1972.pdf. Accessed 25 January 2020.

Smith, Angela and Quaid, Shelia (2017) 'Introduction'. In Smith, Angela (ed.) *Re-Reading Spare Rib*. London: Palgrave Macmillan. Pp. 1–22.

Starkey, Caroline and Tomalin, Emma (2016) 'Building Buddhism in England: The Flourishing of a Minority Faith Heritage'. *Contemporary Buddhism*, 17(2): 326–356.

Stebbins, Robert A. (2005) 'Serious Leisure, Volunteerism and Quality of Life'. In Haworth, John and Veal, Anthony J. (eds.) *Work and Leisure*. London: Routledge. Pp. 200–212.

Sutherland, John (1983) *Offensive Literature: Decensorship in Britain, 1960–1982*. London: Junction Books Ltd.

Varma, Roli (2003) 'E. F. Schumacher: Changing the Paradigm of Bigger Is Better'. *Bulletin of Science, Technology & Society*, 23(2): 114–124.

Veal, Anthony J. (2001) 'Leisure, Culture and Lifestyle'. *Loisir et Société/Society and Leisure*, 24(2): 359–376.

Veal, Anthony J. (2005) 'A Brief History of Work and Its Relationship to Leisure'. In Haworth, John and Veal, Anthony J. (eds.) *Work and Leisure*. London: Routledge. Pp. 15–33.

Veal, Anthony J. (2011) 'Commentary: The Leisure Society I: Myths and Misconceptions, 1960–1979'. *World Leisure Journal*, 53(3): 206–227.

West, Elizabeth (1977) *Hovel in the Hills: An Account of 'The Simple Life'*. London: Faber and Faber.

Woodhead, Linda (2016) 'The Rise of 'No Religion' in Britain: The Emergence of a New Cultural Majority'. *Journal of the British Academy*, 4: 246–261.

Part 3

Belonging

In these final chapters, we turn our attention to the way countercultural migrants to Wales interacted with, and were treated by, the communities they moved to. I wrote and conducted this research in the context of major national debates about immigration reform and attitudes towards migrants. Since this is the context which the research grew out of, and in which the interviews were conducted, it will have impacted how people interpreted and recounted their past experiences. However, migration was a topical issue even at the time. The Windrush generation had only recently arrived in Britain, and the Commonwealth Immigrants Acts of 1962 and 1968, plus the subsequent Immigration Act 1971, were responses to current events. This was the point where contemporary political ideas of immigration as something to be managed and, crucially, of migrants as people who should be integrated, began to come into being. These three chapters each focus on different facets of cultural integration in Wales: the experience of being an outsider, the experience of encountering the Welsh language and positioning oneself in relation to it, and the ongoing impact of the move to Wales on the migrants' identity and sense of belonging.

There is, of course, a big difference between migrants arriving from abroad and the predominantly British countercultural migrants. The differences between 'locals' and 'immigrants' are much less pronounced in the case of internal migrations, where there tend to be fewer variations in ethnicity, religion, and language. It is possible to argue that there is a shared history and lineage between internal migrants and locals, in a way that is difficult with migrants from abroad (McKinlay and McVittie 2007: 175). However, there are still plenty of similarities between the experiences of internal and external migrants. Both go through processes of settling and adjusting into their new communities, and both have to make complex negotiations regarding their new identity and level of integration. Although the majority of the countercultural migrants to Wales were English, and therefore technically classed as internal migrants, this classification disguises the many significant cultural differences between Wales and England. There are long standing tensions between England and Wales due to their historical coloniser/colonised relationship. Because of this, research on Welsh identity frequently draws on the relationship between these two countries (see, for instance, Jones 1992; Jones and Fowler 2007a, 2007b). In chapters which follow, I explore why the move from

DOI: 10.4324/9781003358671-10

England to Wales (or indeed from any other constituent part of the UK) can be just as dramatic in terms of integration as any international move.

At the heart of these chapters is the idea of community, and the way the migrants negotiated their relationship with it. Community, it could be argued, is identity in practice. It is in communities that the identities of multiple groups, including migrants, come together and must learn to coexist (or not). Understandings of what precisely constitutes 'community' have evolved over time, with new meanings developing alongside older ones (Tyler 2006: 22). As mentioned in Chapter 1, early research into communities was dominated by the idea that urban societies were more complex (Tönnies 2012 [1887]: 16). This meant that cities tended to be the focus, a bias which still lingers on today (Valentine 2001: 106–108). In the early twentieth century this focus on cities was combined with an interest in ecology and evolutionary theory, influenced by work from the Chicago School of Human Ecology (Cohen 1985: 22). Community in this era was something understood in scientific terms, measurable and quantifiable. Rural communities were treated with nostalgia, seen to be less advanced than their progressive urban counterparts (Hunter 2018: 10; Entrikin 1980: 49).

In contrast, British sociologists in the 1950s and 1960s associated community with collective identity, focusing on the identities of the working class. Community was understood as a positive thing, something to be valued and aspired towards (Berger et al. 2020: 337–340). In the 1970s, Raymond Williams – who was described as one of the 'founding fathers' of British discourse on community (Horak and Seidl 2010: 1; Berger et al. 2020: 336) – commented on the positivity surrounding 'community', stating that: 'what is most important, perhaps is that unlike all other terms of social organisation (state, nation, society, etc.) it seems never to be used unfavourably, and never to be given any positive opposing or distinguishing term' (1976: 76). This is the traditional understanding of community (Delanty 2003: 35–36). A sense of attachment and belonging is required, but this is seen as proof that 'community' is beneficial, rather than evidence of its potential to exclude (Somerville 2016: 4–5).

A plethora of community studies conducted in the mid-twentieth century helped to cement these two views of community in relation to rural Wales. These included, among others, Rees (1996 [1950]), Frankenberg (1957), Jenkins (1960), Jones (1960), Hughes (1960), Owen (1960), Emmett (1964), Madgwick et al. (1973), and (Cloke et al. 1997: 140). They all drew (consciously or unconsciously) on Tönnies nostalgic view of *gemeinschaft*, portraying Welsh communities as 'vibrant examples of organic development, established over a very long period of time to exist in close harmony with their immediate environment' (Day 2002: 145; Jones 1997: 136). Madgwick, for instance, concluded that Welsh communities were friendlier, closer-knit, and more enthusiastic about participating in community events (Cloke et al. 1997: 142).

More recent research into community takes a very different perspective to these early explorations. Arguments that the previous theorisation implied too great a sense of homogenisation, with little space to tolerate differences, have become more widespread (Smith 1999: 25). Just as feminist and postcolonial scholars argued that identity could be used to exclude as well as include, so they believed could community (Joseph 2002: viii). This darker side of the community was acknowledged in Benedict Anderson's landmark text *Imagined Communities*, which understood community as 'a deep, horizontal comradeship . . . that makes it possible, over the past two centuries, for so many millions of people, not so much to kill, as willingly to die for such limited imaginings' (2006 [1982]: 7). However, Anderson's work has been criticised by postcolonial (and other) theorists for not adequately considering the complexities of multicultural nations such as Wales (Hamilton 2006: 79–80). The idea of community as a form of exclusion was an unconscious part of a great deal of the early research on Welsh communities, which feature recurrent themes of idyllic Welsh rural communities 'under attack' by anglicisation (Day 1998: 240). Lynn Staeheli summed up this contrasting image of community by arguing that

> [C]ommunity is where contests are waged over membership and the political subjects and subjectivities that 'belong' in a political community. Community is contested, struggled over. Sometimes this leads to a radical openness. Sometimes community is subject to strict control and discipline.
>
> (2008: 7)

The vast majority of the studies discussed treat community as something that is territorially defined. They were far from unique in doing so – 'neighbourhood' is often used interchangeably with 'community' (Valentine 2001: 121) – but communities do not always develop around specific places. Over time, as people's social relationships become more dispersed and less dependent on physical proximity, communities developed around shared identities or lifestyles are becoming more common (Wills 2012: 116). This is not always properly recognised in the literature discussed earlier, as community is still frequently theorised around the idea of territorial boundaries (Herbert 2005: 853). In this book, I use 'community' to refer to both the interactions between people in local areas (the 'identities in practice' referred to earlier) and groups which share similar interests but are not necessarily close in proximity (such as the wider alternative community).

In the following chapters, I try to build on the developments in the way community is constructed and understood, recognising its flaws as well as its strengths. The choice to do this through stories from migrants themselves is in itself part of this research lineage as, just as in the case of identity research discussed in Part 2, shared narratives have come to be seen as essential to conceptions of community. As Rappaport argued at the turn of the century,

> the psychological sense of community can be indexed by its shared stories. People who hold common stories about where they come from, who they are, and who they will, or want to be, are a community. A community cannot be a community without a shared narrative.
>
> (2000: 6)

References

Anderson, Benedict (2006 [1982]) *Imagined Communities: Reflections on the Origin and Spread of Nationalism*. London: Verso.

Berger, Stefan, Dicks, Bella and Fontaine, Marion (2020) '"Community": A Useful Concept in Heritage Studies?'. *International Journal of Heritage Studies*, 26(4): 325–351.

Cloke, Paul J., Goodwin, Mark and Milbourne, Paul (1997) *Rural Wales: Community and Marginalization*. Cardiff: University of Wales.

Cohen, Anthony P. (1985) *The Symbolic Construction of Community*. Chichester: Ellis Horwood.

Day, Graham (1998) 'A Community of Communities? Similarity and Difference in Welsh Rural Community Studies'. *The Economic and Social Review*, 22(3): 233–257.

Day, Graham (2002) *Making Sense of Wales*. Cardiff: University of Wales Press.

Delanty, Gerard (2003) *Community*. Abingdon: Routledge.

Emmett, I. (1964) *A North Wales Village: A Social Anthropological Study*. London: Routledge.

Entrikin, J. Nicholas (1980) 'Robert Park's Human Ecology and Human Geography'. *Annals of the Association of American Geographers*, 70(1): 43–58.

Frankenberg, R. (1957) *Village on the Border*. London: Cohen and West.

Hamilton, Mark (2006) 'New Imaginings: The Legacy of Benedict Anderson and Alternative Engagements of Nationalism'. *Studies in Ethnicity and Nationalism*, 6(3): 73–89.

Herbert, Steve (2005) 'The Trapdoor of Community'. *Annals of the Association of American Geographers*, 95(4): 850–865.

Horak, Roman and Seidl, Monika (2010) 'Raymond Williams: Towards Cultural Materialism: An Introduction'. In Seidl, Monika, Horak, Roman and Grossberg, Lawrence (eds.) *About Raymond Williams*. Abingdon: Routledge. Pp. 1–17.

Hughes, T. (1960) 'The Social Geography of a Small Region in the Llyn * Peninsula'. In Davies, E. and Rees, A. (eds.) *Welsh Rural Communities*. Cardiff: University of Wales Press.

Hunter, Albert (2018) 'Conceptualizing Community'. In Cnaan, Ram A. and Milofsky, Carl (eds.) *Handbook of Community Movements and Local Organisations in the 21st Century*. Springer.

Jenkins, D. (1960) '"Aberporth": A Study of a Coastal Village in South Cardiganshire'. In Davies, E. and Rees, A. (eds.) *Welsh Rural Communities*. Cardiff: University of Wales Press.

Jones, E. (1960) 'Tregaron: The Sociology of a Market Town in Central Cardiganshire'. In Davies, E. and Rees, A. (eds.) *Welsh Rural Communities*. Cardiff: University of Wales Press.

Jones, Noragh (1997) 'Diverging Voices in a Rural Welsh Community'. In Milbourne, Paul (ed.) *Revealing Rural 'Others': Representation, Power and Identity in the British Countryside*. London: Pinter. Pp. 135–146.

Jones, R. Merfyn (1992) 'Beyond Identity? The Reconstruction of the Welsh'. *Journal of British Studies*, 31(4): 330–357.

Jones, Rhys and Fowler, Carwyn (2007a) 'National Élites, National Masses: Oral History and the (Re)production of the Welsh Nation'. *Social & Cultural Geography*, 8(3): 417–432.

Jones, Rhys and Fowler, Carwyn (2007b) 'Where Is Wales? Narrating the Territories and Borders of the Welsh Linguistic Nation'. *Regional Studies*, 41(1): 89–101.

Joseph, Miranda (2002) *Against the Romance of Community*. Minneapolis: University of Minnesota Press.

Madgwick, P., Griffiths, N. and Walker, V. (1973) *The Politics of Rural Wales: A Study of Cardiganshire*. London: Hutchinson.

McKinlay, Andy and McVittie, Chris (2007) 'Locals, Incomers and Intra-National Migration: Place-Identities and a Scottish Island'. *British Journal of Social Psychology*, 46: 171–190.

Owen, T. (1960) 'Chapel and Community in Glanllyn, Merioneth'. In Davies, E. and Rees, A. (eds.) *Welsh Rural Communities*. Cardiff: University of Wales Press.

Rappaport, Julian (2000) 'Community Narratives: Tales of Terror and Joy'. *American Journal of Community Psychology*, 28(1): 1–24.

Rees, A. (1996 [1950]) *Life in a Welsh Countryside*. Cardiff: University of Wales Press.

Smith, D. M. (1999) 'Geography, Community, and Morality'. *Environment and Planning A*, 31: 19–35.

Somerville, Peter (2016) *Understanding Community: Politics, Policy and Practice*. Bristol: Policy Press.

Staeheli, Lynn A. (2008) 'Citizenship and the Problem of Community'. *Political Geography*, 27: 5–21.

Tönnies, Ferdinand (2012 [1887]) 'Community and Society'. In Lin, Jan and Mele, Christopher (eds.) *The Urban Sociology Reader*. Abingdon: Routledge.

Tyler, Richard (2006) 'Comprehending Community'. In Herbrechter, Stefan and Higgins, Michael (eds.) *Returning (to) Communities: Theory, Culture and Political Practice of the Communal*. Pp. 21–28.

Valentine, Gill (2001) *Social Geographies: Space and Society*. Harlow: Prentice Hall.

Williams, Raymond (1976) *Keywords: A Vocabulary of Culture and Society*. Oxford: Oxford University Press.

Wills, Jane (2012) 'The Geography of Community and Political Organisation in London Today'. *Political Geography*: 114–126.

7 Fitting into a new community

We have already touched on some of the ways that our group of migrants interacted with the communities around their new homes. We know that they interacted on an economic level, starting local businesses, that they participated in social organisations, and that they joined a wide variety of spiritual groups. But, we have heard little, thus far, about the way they settled into their new communities. How did they interact with their new neighbours? Were they accepted or treated as outsiders? This chapter is where we explore the migrants' unfolding relationship with those around them, and the way they integrated themselves into their new communities.

Research into migrant integration is often heavily influenced by the idea that there is a binary distinction between 'locals' and 'migrants', but this distinction has been widely disputed for being too simplistic (Jedrej and Nuttall 1996: 173). When considering the integration experiences of countercultural migrants we need to consider exactly who they are integrating with; it is just as important and valid to integrate into the existing community of countercultural migrants, for instance, as it is to integrate with neighbours who are locally-born. I generally use the term 'local' to refer to members of the host community, that is, anyone already resident in the area who is not part of the countercultural community. There are of course overlaps, and I shall endeavour to try and make it as clear as possible, when recounting migrants' stories, who they are forming a community with.

7.1 'A space in-between'

Writing in the context of migration to remote Scottish islands, sociologists Kathryn Burnett and Lynda Stalker found that

> in-migrants to Scotland's islands, like many who move to rural and remote areas, must negotiate a claim to place. Moving to an island often requires a figuring out where you fit in and how to define your identity in relation to island space and to 'islandness'.
>
> (2016: 239)

The same negotiation of a claim to place was required from migrants to rural Wales, who found themselves existing within pre-existing communities permeated

DOI: 10.4324/9781003358671-11

by 'Welshness'. They had to find a space for themselves within these new communities, a way of positioning themselves in relation to the existing social structures.

Carol arrived in Wales from Manchester as a teenager in 1967. She had moved with her parents, who felt, like so many others, that they needed to get away from the stresses of their previous lives. In order to earn some money the family (Carol was the eldest of four children) rented a pub in a small village in Carmarthenshire, not far from the Brecon Beacons. Unlike some migrants, they didn't know anyone in the area they moved to or indeed anyone else who had moved to Wales at all. They were negotiating their place in their new community completely from scratch. Because they were taking over the pub, traditionally a central hub of village life, they suddenly found themselves playing an integral role in an entirely new and unfamiliar community and culture. Admittedly, as was so often the case for our migrants, there was some work to be done first:

> We rented a run-down pub that had been empty for years and years and years, a room with a bar which people had just come in and opened up and run and then shut at the end of the night and gone away again. Um, I mean it even had a tree growing in the corner of the living room when we bought it, well we rented it. So, we had to spend quite a lot doing it up.

But, alongside the work to be done on the pub was the work that needed to be done to fit into the new world they had moved to. They were the only English family in the village; hence, they were just as strange to the locals as the locals were to them. To some extent, this could be an advantage: 'Even coming to the pub, people used to do it as a visit to the slightly strange family, more than just coming for the beer'. The family were seen as being unusual, but they didn't encounter outright hostility for being English. Carol recalled how:

> I don't remember there being any anti-English feeling when we were there, which is quite interesting. . . . I don't remember that. As I said, being treated as a bit of a freak and an oddity isn't the same thing as being resented because you're English. I don't ever remember anything like that and I don't remember anybody ever saying anything.

On the whole, they were accepted very well by the local community:

> The local smugglers were coming in, they used to leave us salmon at the back, wrapped in paper on the back step regularly, I remember that. They used to come in and they used to sort of nudge you behind the bar and say 'if you go out to the empties you might find something there.

For Carol, the experience of settling in Wales was one of being the odd family out, the only English in the village, not only of quite fully understanding the culture and community at first but also of becoming part of the community at the same time – still different, but accepted.

For Barbara, the experience of negotiating a different culture was not new. As we learned in Chapter 1, she had previously left her native Ireland to live in France and London before arriving in Wales. She found that being Irish gave her an advantage over other incomers in the eyes of the locals, recalling that:

> I don't think I ever had any difficulty in being accepted, partly because I was Irish. . . . I think I was immediately accepted, you know, you're Irish it's like the same as being Welsh almost, you know'. Marko Valenta, in his study of immigrant identity, found that 'people artfully engage in 'biographical work' to assemble aspects of personal history that can be used to support present claims about themselves.
>
> (2009: 364)

In this way, her awareness of the similarities between 'Irishness' and 'Welshness' helped Barbara negotiate and settle into the local community.

Other things also helped with this process of fitting in. Barbara became involved in the community on a variety of levels, mixing with incomers and locals:

> I also, I suppose, became involved in the community to a certain extent. Again, I think probably my, my rural Irish background, you just had to know what everybody was doing and everybody had to know what you were doing. . . . Very similar here. Um, so I used to go, there'd be a circle of coffee mornings where we went to people's houses, English incomers and Welsh. . . . Grew my own vegetables and um, yes became more involved in self-sufficiency things. And met people like Dave Frost, who ran the organic farm foods. And um, yes, I suppose got to know the locals a bit better, became a little more integrated.

When Barbara arrived in Wales, she was single, but in her time here, she married another incomer and had four children. Although this naturally happened over a longer timescale than some of the others' stories, the arrival of her children was crucial to Barbara's integration into the community:

> When I wanted to take my younger children to playschool what happened was just that very week they were losing their teacher, in the playschool in Llanbadarn. And they said 'do you know anyone who's got teaching qualifications?', and I said 'well I have as well', 'so would you like to run our playgroup?' Ok, well I wasn't really thinking about that because the son I was taking to playgroup was three, and my baby was nine months old. So I thought I was a bit young. I said 'well I've got a young baby who isn't play-school age', 'oh you can bring him with you if you please please please run our playgroup', so I said ok.

Barbara's involvement with local schools and playgroups was central to her early experiences of integration. We'll discuss the significance of children to migration more fully later on, but given its significance it is worth acknowledging here too. Many of the migrants described the importance of their children to the early stages of integration. Enrolling children into a local school gives parents a ready-made way of interacting with other local parents, and the opportunity to become further involved through after-school activities, parent-teacher associations, etc. School brings incomers and locals together in a way that few other organisations can. Rural schools are proven to improve the vitality of their communities, and healthy engagement between the two provides social and economic benefits (Casto 2016: 143). Because schools are a necessary part of local communities, Barbara felt that she was able to provide a service rather than impose:

> I sort of fell into a space. A space like in between, 'cause they didn't have anybody and they needed somebody. So again, I became fairly involved with local mothers, I set up a mother and toddler group as well, and um, sort of just got to know a lot of people in the area, and a lot of local Welsh people. I ran the playgroup for quite a few years.

This idea of falling into space, filling existing gaps in the community, is central to Barbara's story of integration. She took on roles in the community that were not being taken by locals and lived in homes that had been left derelict, always finding ways to augment the community rather than act in competition with it:

> I'd always have felt wary of moving in and buying a house that indigenous people had wanted, but nobody had wanted this. The same as nobody wanted to work in the playgroup. So I always feel that a history of my story is falling into the bits in between, you know, and going with it.

Taking on voluntary roles was central to Barbara's experience of fitting in. In rural communities, these sorts of roles are most commonly taken on by women (Hughes 1997: 178). As noted by Jo Little (1997), they are common and effective ways of integrating into rural communities. But, Little also notes that this may not always be a positive thing, as rural communities could be seen as taking advantage of women's unpaid labour by relying on their voluntary contributions (203–205). The people I interviewed presented their volunteer work in a positive light, but this is not always the case for everyone.

The idea of filling the 'spaces in between' is a recurring one. It is one of the ways people dealt with encountering 'Welshness'. Although Barbara recalled a few difficulties in negotiating 'Welshness' alongside 'Irishness', being English could be more complicated. The idea of integrating to a level beyond 'sitting on the surface' was a way of coping – just as we saw with Carol and her family's involvement with the local pub. The process of finding one's place in a new community, negotiating fluid landscapes of Welshness, happens through everyday activities, practices, and social interactions (Antonsich and Matejskova 2015: 503). Communities, on both local and national scales, do not simply exist; they must be continuously re-made and re-affirmed through a multitude of different practices – some mundane and some less so (Jones and Fowler 2007b: 90). The everyday lives of migrants once they arrived in Wales will have affected how they were received, varying according to place and community, but always connected. Everyday activities can also disrupt belonging rather than reinforce it (Antonsich 2018: 456). In this way, the migrants were not only encountering a new set of cultures and identities but also actively working to create or disrupt them and to affirm their place within them.

Let us take Carol's story as an example. Having taken on the role of local pub owners, Carol and her family automatically acquired an important position within the village. They integrated into local practices, making sure they were open at the appropriate times and serving what the locals wanted to purchase. They accepted the local poachers' gifts of salmon, fitting themselves into the community rather than disrupting it by reporting them. But at the same time, they still remained separate. None of the family learnt to speak Welsh fluently, and while they were not met with hostility for this, it could still be uncomfortable, particularly given their occupations:

> Everybody spoke Welsh in the pub. So they would speak English to you, and then, they would carry on. Which did mean that frankly you were excluded, but it was, I never felt it was intended to exclude you. It was just obviously that's the natural thing for them to do. But it did mean you didn't always know what was going on, which can be a little bit frustrating. It could be a bit boring sometimes, you know.

This excerpt illustrates how being accepted into the community on one level did not mean being treated exactly as a local (if there is even such thing as a standard 'local' to compare with). Even with their prominent position in the community, the family found themselves happier when Irish workers began arriving to work on the construction of the Llyn Brianne dam, find that it was easier to get along with other people who weren't locals:

> They were great fun, and they used to like a drink, and they used to like a laugh, and of course they were all away from home. They were all, most of them, young and fit, so you know, it was quite nice, a bit of fresh air in the village I think, . . . I think after that, 'cause they didn't stay that long, I can't remember how long they took to build it, I think it was two or three years, and then after that they went. And I feel like it was very quiet.

Carol and her family could be at home on some levels and still separate on others. They could be simultaneously accepted and excluded. These aren't binary forms of identity; they exist alongside each other, all part of the same story. Josie observed how 'there were lots of pockets of different tribes really, 'cause it felt quite tribal. But we would all get together at different times'. For her, the bonding came from festivals, particularly Meigan Fayre, which was held near Cardigan. Although it is different in many ways, this is similar to Carol's experiences of bonding with the community through the pub, while simultaneously being part of separate groups – or, as Josie calls them, 'tribes'.

7.2 Social capital and integration

The activities that were central to Carol and Barbara's early integration experiences, running community groups and services, were not just ways of finding 'a space in between'; both can also be seen as ways of exchanging social capital. Such exchanges of social capital are thought to be an important part of migrant integration (Wessendorf and Phillimore 2019: 125–126). It usually comes in the form of knowledge about the migrants' new homes; Kennan et al. have argued that 'information is the most critical need of new settlers', and when migrants cannot access the 'information landscape' of their new home, they risk social exclusion (2011: 192–194). In Wales, the kind of information needed by settling migrants is very different. In other countries, language is usually central – migrants generally place particular value on knowledge relating to their new local culture and language – but of course in the 1970s Wales, there was no real need to learn Welsh in order to function. Welsh may have been a central part of the cultural difference and accompanying tension, and this is something that we will return to later on, but English was still the official language. This means that language knowledge, usually highly valued among migrants, becomes less important to incomers to Wales. Likewise, although there are significant cultural differences between England and Wales, the culture shock for people moving between the two is significantly less than for an English person moving to, say, Afghanistan. Hence, cultural knowledge, although still important, is also less sought after than in most cases.

This does not mean that the theory of social capital exchange is irrelevant to countercultural migrants, just that the knowledge being exchanged takes a very different form than usual. Since many of them moved in order to adopt more environmentally friendly, self-sufficient lives – and, crucially, they did so with little prior experience of running a smallholding – information about the practical side of their new life was highly prized. For this reason, many key early interactions, with both 'locals' and earlier migrants, were focused on sharing farming knowledge, ideas, experience, and tools. Freya, George, and Gillian found that official information services and advice providers were essential to their process of settling into life in Wales. This means that structured forms of interaction were integral to the early days:

Freya: We contacted, at the time there was something called ADAS.
Gillian: Yeah, I think so.
Freya: Agricultural-
George: Department Advisory Service.
Freya: There you go.
Gillian: What was it?
George: Agricultural Department Advisory Service, something like that.
Freya: It certainly ended 'advisory service', yeah. And this man came and he didn't, um-
George: He was really good wasn't he?
Freya: He was so nice, and he didn't diss us at all, you know, he didn't sort of 'you've gotten acres what are you doing?', you know. And um, but he was the one that got our water tested, got our soil tested.

George: Yeah.

Freya: So you know, all what it needed and gave us, when I was saying 'well my potatoes are scabby' and he sent leaflets about what you could do about that. And um, free service then, it was, yeah.

Gillian: It was also things like lambing classes, again free, you could go to a learn how to pull a lamb, you know.

Knowledge like this was highly valued among countercultural migrants, and essential to a smooth settling-in period. The Agricultural Department Advisory Service (ADAS) had been created in 1972, as an amalgamation of the previous National Agricultural Advice Service and Land Service. It was a non-elected quango, with responsibilities for various issues including drainage, woodlands, hedgerow management, etc. (Wilson 1998: 29; Winter 1996: 246). Although established with the intention of supporting larger-scale, commercial farms, ADAS proved to be a lifeline for inexperienced smallholders. John Seymour's books were able to inspire them and keep them informed of the theory, but they could not provide the personalised advice and training that migrants needed when settling themselves in. Jim, who we met in Chapter 2, recalled that 'I think lots of people would like to wring [Seymour's] neck now, because he obviously made his living writing books rather than being self-sufficient!'

Regardless of what one's opinion on John Seymour may have been, the ability to see practical advice demonstrated first-hand, and to be able to talk things over with someone who had been there before, was invaluable to anyone involved in agriculture. Facilities such as ADAS were particularly useful because, since they were intended for both locals and incomers, they were a way for the two groups to connect on an equal footing. As we will discuss later on, language classes put learners below speakers in the social hierarchy, but ADAS bypassed this division. Sometimes, the transmission of knowledge went the other way, and migrants were able to integrate by sharing their own experiences. Lesley described earlier how one of her first experiences in Wales was having her car hauled out of a ditch by the local farmer. This farmer went on to be one of her key links to the local community:

> When he found that I'd actually been to horticultural college and I could drive a tractor and all these things, within a week I was helping out on the farm. So that was great, that was a really good . . . 'cause historically there's always been this link, I'm something of a historian, this link between smallholders and the sort of larger farms, a sort of symbiotic link. So yeah, I was soon helping out on the farm, and there was never any question of any pay or anything like that, I would just find a sack of potatoes or a bale of hay or straw dumped outside the house. And that was really fortunate, he didn't have any children and I sort of became honorary farmer's son. Yeah, so that was good. I mean apart from that it was, it was a very isolated way of life, particularly for me, um, 'cause I wasn't going out to work and it was literally in the middle of a field.

Lesley remembered her early days in Wales primarily as lonely and isolated, but in addition to making friends with the nearby farm she was also able to form connections with local goatkeepers. These were all more personal connections than Freya, George, and Gillian's passing acquaintanceship with the ADAS man. Susanne Wessendorf and Jenny Phillimore argue that integrating migrants form different levels of relationship, each of which plays a different role in the integration process. There are 'fleeting encounters' which are crucial to the settlement process but don't necessarily lead to anything deeper. Then there are 'crucial acquaintances', relationships formed through work or other organisations, like the ADAS man. Finally, there are true friendships (2019: 124). We saw in Lesley's case how a 'crucial acquaintanceship' with the local farmer was able to become a 'true friendship' with time. The exchange of knowledge/social capital provided a useful catalyst for the development of these relationships. It helped, too, that the countercultural lifestyle benefitted locals – it contributed to the local agricultural economy, and helped to keep old traditions and ways of life alive. Because of this, Lesley's response when asked if she felt tension with the locals was:

> Not at all, which is interesting when you think that this is not that far off the point where holiday homes were being burnt. But of course we weren't, we were actually moving there, into a pretty deserted hamlet. . . . And in a way it was a way of sort of . . . it's like the local school wouldn't have kept going without the incomers. . . . We sort of, it was, it revived the communities.

7.3 Children's experiences

As we saw in the aforementioned stories, involvement in schools, playgroups, and parent groups was an effective way for adults to develop relationships with locals and other migrants. This was especially the case for women; involvement in children's organisations is often regarded as a particularly important part of women's participation in rural communities (Little and Austin 1996: 105–106). But what about the experiences of the children themselves? For those who were born in Wales 'integration' was a non-issue: as we'll discover later, most of these children came to consider themselves Welsh. For those that moved with their parents though, the situation was very different. Moving as a child is a hugely disruptive experience, very different to the process of moving as an adult. Children are subjected to a different range of stressors, find themselves placed in a different kind of society, and frequently have little proper understanding of why they are being made to uproot themselves. They participate in exchanges of social capital in completely different ways to adults. I was fortunate enough to interview several people who moved to Wales as children, and it is important to consider their side of the story as well.

We first met Judith, who moved to Wales with her husband and children in 1974, back in Chapter 1. Judith's reasons for moving to Wales were twofold: she wanted to support her husband and his work on cooperative living, and she wanted to get her family away from the dangers of city life. Judith's daughter Catherine was only five at the time, so naturally her impression of moving to Wales was very different. The age at which children move has a big impact on how well they adjust and integrate. Young children usually have less trouble adjusting to a new culture than older children, but this does not necessarily mean that Catherine had it easy (Berry 1997: 21–22). Too young to truly understand the reasons behind the move, or recognise any problem with living in the city, Catherine was initially reluctant to embrace the shift from city to rural living:

> I remember the multicultural aspect of living in London. And I know I liked school, 'cause my teacher was from Australia. Um, I had an Indian friend called Pooja, and there was mum's friend's daughter called Candice. Um, there were people from all over the world and um, yeah, I liked the multicultural aspect of London. Then I think that for me, leaving, I wasn't really aware that we were leaving London. Um, I remember a holiday in Cornwall. I remember um, Mum had given us a baby, my sister Vicky. And um, Vicky was about six-months old I think when we moved to Barmouth. And then I remember Mum taking me to school and um, in a blue coat that I had, with the red zip. And um I, I think I announced to people that I was only going to be there for a year. I think I made that quite clear, that I was from London and I was going back.

For adults, schools were useful ways of integrating with others in the community. For children, they take on an extra level of pressure and significance. Adults can retain a level of control over their involvement with schools, they can take a step back if they feel unwelcome, but children do not usually have this option. Because of this, any cultural difference is imposed upon children in a way that it is not for adults. Catherine recalls the contrast between her multicultural London school and much more traditional Barmouth school as one of the biggest shocks of arriving in Wales:

> One of the first things that hit me was the alphabet. And I, I said to them 'no that's not, that's not how you say the alphabet. The alphabet is ABC [English pronunciation]'. And they said 'no it isn't, it's ABC [Welsh pronunciation]', and I went 'no, you're wrong'. So, there was that. And then, there was the whole Christianity aspect, so um, you know, if you go to school in um, in Wales they assume that you're a Christian, and um, we had to say prayers. So, I'd never had that before.

Neither of these are situations that an adult would be likely to encounter without some action on their part; one would have to sign up for a Welsh lesson for instance or choose to attend a religious service. But children going to school were automatically placed in these situations without the chance to opt out. The question of whether to learn Welsh was one of the most controversial issues faced by migrants to Wales. In addition, many families enrolled their children into completely Welsh language schools, either in an attempt to integrate or simply because it was the nearest and most convenient option (Barbara, as we saw earlier, was one of this group, and found it very helpful for her own integration). Virtually all children of migrants therefore had to learn Welsh up to a certain level, regardless of what their parents did.

For Catherine, the cultural adjustments did not end with Welsh classes and religious observations:

> The other aspect was there weren't any other, um, children from other places. So they were all white faces and um, I missed the multicultural aspect of having kids from all over the world. And um, there just seemed to be a lot of control, and a lot of restrictions, not so much freedom, and um, there was an overwhelming smell, the smell that you would get in a caretaker cupboard. You know, of old mop and disinfectant and sick, combined together. And I remember thinking, 'oh my God, this smell is nauseating!'. But we did have very nicely polished floors! So um, I wasn't very happy. And it was the beginning of my experience of being the outsider. And um, there was definitely a sense of um, Welsh kids go to Welsh class and English kids go to the English class and um, it was a very different um, attitude towards education in 1974, because they kept us separate.

This last observation that Welsh children and English children were treated as different and separated is a particularly interesting one. For Catherine, going to school meant not just being automatically immersed into a new cultural context but also having her place in this context predetermined. It is hardly surprising that this was when she began to feel like an outsider. This is a recognised feature of migration: as well as bringing their own identities, migrants have identities ascribed to them by locals. They must either accept or reject and reconstruct these preconceptions; challenging enough for an adult, never mind a child (Valenta 2009: 352). The Welsh language was a key part of this: migrants were ascribed the identity of 'non-Welsh speaker' and automatically segregated accordingly. Language learning can have many positive effects on integration but it can also be used to isolate, as we see happening here. It is possible to be well-integrated in theory and appearance, but still feel lonely and disconnected socially (Wessendorf and Phillimore 2019: 133). School is an example of this: the children of migrants were interacting daily with locals and learning a great deal about Welsh culture and language, but without the acceptance of their peers, schoolchildren remain disconnected. It is a microcosm of the wider paradox of migration identified by Burnett and Stalker: migrants are seen as perpetually excluded from local communities, while simultaneously expected to integrate and engage with local communities' customs and traditions (2016: 252).

Gwyneth was another countercultural migrant who moved as a child with her parents. She was older than Catherine, having arrived in Wales at age 10, narrowly avoiding having to take the dreaded eleven-plus exam at her previous school in Gloucestershire. Although born and brought up in England, with an English mother and younger sister, Gwyneth's father Philip was Welsh. Philip had left his childhood home in south Wales to work as a teacher in London during the 1950s, but in 1965, the family uprooted to Harlech to be closer to his parents. It may seem that having a family connection to Wales should have made the adjustment easier for Gwyneth, but this was not the case. Philip had come from a very different part of

Wales, the mining valleys of South Wales, so Gwyneth still felt a strong sense of displacement despite her second-hand knowledge:

> The main difference I noticed was that the school, some of the lessons were in Welsh. Um, so for, in the primary school. In the secondary school they were mostly in English and they were, it, the Welsh was set to one side. Welsh, true Welsh speakers actually had different lessons. I think it was geography and history, pretty certain, and I think music, but then that's, that were taught in Welsh. . . . So whenever those lessons were on the timetable they just sent me and Fergus Mercer out of the room with a pile of encyclopaedias.

Like Catherine, Gwyneth found herself automatically assigned a place in the new culture that she was suddenly immersed in. Gwyneth did try to learn Welsh and found that the skills she managed to pick up helped to some extent: she could at least understand the school assemblies now if nothing else. But the separation still lingered on:

> My friends at school were all basically English speakers who had moved here. There was almost a divide. Just culturally among the people that had always lived here, and the people that . . . import, if you like. People who hadn't. And I would, much as I would have, I was just never going to fit in with them because I couldn't speak their language. Not at the speed they did.

It is a stark contrast with what Gwyneth's father experienced re-settling in Wales. Philip did not speak Welsh either, as the South Wales valleys were predominantly English speaking even then but never found it caused much of a problem in Harlech:

> I always felt at home in Wales. This is where I belong, so to speak. Although I didn't speak Welsh at all. Um, but I, I think it was more a trial for Gwyneth than it was for me. I mean people tended to speak in English to me anyway.

7.4 The impact of anglicisation fears

Two narratives have tended to dominate discussions of migration into rural areas. These can be broadly generalised as one positive, and one negative. Burnett and Stalker argue that there is a dominant negative narrative of the impact migration (particularly internal migration) has on rural communities. House price inflation, gentrification, the decline of minority languages, and loss of traditional culture are all seen as the characteristic results of migration. But, as Burnett and Stalker state, this narrative oversimplifies the experiences of both migrant and rural communities (2016: 241). The other perspective is that exemplified by Barbara and Carol's stories, of incomers benefiting local communities by providing needed services and contributing to the local economy. Immigration has been found to provide the receiving community with: 'fiscal advantages, increased gross domestic product per head, a ready supply of labour, and improvements to the age structure' (Coleman and Rowthorn 2004: 579).

As discussed in Chapter 2, in-migration to Wales was being actively encouraged by local councils at this time, in order to counter the effects of out-migration. The nationalist backlash against English incomers did not truly take off until the wave of second-home ownership began in the 1980s, leading Meibion Glyndwr (Sons of Glyndwr) to launch their famous arson campaign (Day et al. 2010: 1407). When the countercultural immigrants discussed here were arriving, tensions had not yet peaked, although they were mounting. An aversion towards English incomers had been around for years, Wales is a colonised nation after all, and fears of 'anglicisation' had been voiced since at least the nineteenth century. A succession of events had led to a greater English presence in Wales, the construction of the railways for one, evacuation during World War Two another, and countercultural incomers were just the latest development (ibid: 1406). Hence, there was a real-life tension between the desire to benefit from the arrival of incomers, and the worry over an English threat to Welsh culture and way of life.

In Wales, it is the divide between Welsh and English which is most significant, and this is where the lines between national identities are most fiercely drawn (Day et al. 2010: 1408). In the 1970s, this tension was most clearly exemplified in the campaign for Welsh devolution. Given its significance, it is worth taking a moment to give an overview of this campaign. Devolution was something that had been in the minds of Welsh politicians for a long time prior to the 1970s, spoken about since at least the 1880s (Jones and Scully 2003: 86). Nationalism was on the rise in Wales during the 1960s, with the popularity and membership of Plaid Cymru growing dramatically, and Cymdeithas yr Iaith Cymraeg (the Welsh Language Society) established in 1962 (Rawkins 1978: 525; Thomas 1997: 326). It was not just Welsh nationalists calling for a referendum, with demands also being made by Labour politicians in Wales (Evans 2006: 150–151). Although the campaign was ultimately unsuccessful, with devolution rejected by a ratio of four to one in the 1979 referendum, it was still a significant part of Welsh politics in this era (Jones and Scully 2003: 88).

Despite its significance, none of the project contributors spoke specifically about devolution, although it was happening right when they were arriving and living in Wales. In a paper published in 1978, Phillip M. Rawkins argued that Welsh nationalism was an attempt 'to make sense – politically and psychologically – of the experience of life on the periphery of the economically advanced core of the modern world-system' (519). This observation is interesting, as it suggests that there is some parallel between Welsh nationalism, including the campaign for devolution, and the countercultural movement. The counterculture was also, ultimately, about finding one's place in the world. It did this by rejecting aspects of mainstream culture. Welsh nationalism tried to achieve the same ultimate goal on a national scale, by rejecting English control. One of its triggers was as a reaction to English presence in Wales – which makes it even more interesting that none of the migrants mentioned it.

Susan Pitchford argued that Wales has been more successful than England at maintaining strong communities. Because of this, the promise of community is one of the things that draws people from England to Wales (2001: 58). This is borne out by the interview data, which, as we saw in Chapter 1, shows that many people moved away from cities due to the lack of community and connection they found there. However, once they arrived, they could be seen as a threat by Welsh nationalists to the communities they moved to join (ibid). Negative stereotypes of the English abound in Wales. They were thought to be arrogant and privileged, taking over jobs and imposing their lifestyle on locals, or 'social misfits' that were unwanted back in England (Day et al. 2010: 1408). Our particular group of countercultural migrants seems to have encountered relatively little animosity in this respect, which possibly explains their lack of reference to devolution.

Welsh historian Martin Johnes noted that very little work on the history of Wales has been written from the perspective of the outsider. Since most historians of Wales tend to be Welsh themselves, their work is often bound up with an insider's view of Wales and Welshness (Johnes 2010). We have seen a different perspective here, a story which has not often been told, which highlights the development of people's lifestyle, identity, and community role in a different way. We've learnt how these migrants fitted into their new communities by finding the 'spaces in between' – ways of slotting into existing roles without impinging on those who were already there. We have learnt how exchanges of social capital were also important – although not always in the way one would expect for migrants to a new country. Children's experiences differed from adults, thanks to an added layer of expectations and control placed upon them, and this could exacerbate the division between migrant children and locals. We have heard how fears over anglicisation were growing as a result of the incomers, but the stories we have heard so far have been relatively positive. It is in the next chapter, when we turn our attention towards the role of the Welsh language in these migrants' lives, that the extent of the tensions becomes truly apparent.

References

Antonsich, Marco (2018) 'The Face of the Nation: Troubling the Sameness-Strangeness Divide in the Age of Migration'. *Transactions of the Institute of British Geographers*, 43(3): 449–461.

Antonsich, Marco and Matejskova, Tatiana (2015) 'Immigration Societies and the Question of "The National"'. *Ethnicities*, 15(4): 495–508.

Berry, John W. (1997) 'Immigration, Acculturation, and Adaptation'. *Applied Psychology: An International Review*, 46(1): 5–68.

Burnett, Kathryn A. and Stalker, Lynda Harling (2016) '"Shut Up for Five Years": Locating Narratives of Cultural Workers in Scotland's Islands'. *Sociologia Ruralis*, 58(2): 239–257.

Casto, Hope G. (2016) '"Just One More Thing I Have to Do": School-Community Partnerships'. *School Community Journal*, 26(1): 139–162.

Coleman, David and Rowthorn, Robert (2004) 'The Economic Effects of Immigration into the United Kingdom'. *Population and Development Review*, 10(4): 579–624.

Day, Graham, Davis, Howard and Drakakis-Smith, Angela (2010) '"There's One Shop You Don't Go into If You Are English": The Social and Political Integration of English Migrants into Wales'. *Journal of Ethnic and Migration Studies*, 36(9): 1405–1423.

Evans, John Gilbert (2006) *Devolution in Wales: Claims and Responses, 1937–1979*. Cardiff: University of Wales Press.

Hughes, Annie (1997) 'Women and Rurality: Gendered Experiences of "Community" in Village Life'. In Milbourne, Paul (ed.) *Revealing Rural 'Others': Representation, Power and Identity in the British Countryside*. London: Pinter. Pp. 167–188.

Jedrej, Charles and Nuttall, Mark (1996) *White Settlers: The Impact of Rural Repopulation in Scotland.* Luxembourg: Harwood Academic Publishers.

Johnes, Martin (2010) 'For Class and Nation: Dominant Trends in the Historiography of Twentieth-Century Wales'. *History Compass*, 8(11): 1257–1274.

Jones, Rhys and Fowler, Carwyn (2007b) 'Where Is Wales? Narrating the Territories and Borders of the Welsh Linguistic Nation'. *Regional Studies*, 41(1): 89–101.

Jones, Richard Wyn and Scully, Roger (2003) 'A "Settling Will"? Public Attitudes to Devolution in Wales'. *British Elections & Parties Review*, 13(1): 86–106.

Kennan, Mary Anne, Lloyd, Annemaree, Qayyum, Asim and Thompson, Kim (2011) 'Settling in: The Relationship between Information and Social Exclusion'. *Australian Academic & Research Libraries*, 42(3): 191–210.

Little, Jo (1997) 'Constructions of Rural Women's Voluntary Work'. *Gender, Place and Culture*, 4(2): 197–210.

Little, Jo and Austin, Patricia (1996) 'Women and the Rural Idyll'. *Journal of Rural Studies*, 12(2): 101–111.

Pitchford, Susan R. (2001) 'Image-Making Movements: Welsh Nationalism and Stereotype Transformation'. *Sociological Perspectives*, 44(1): 45–65.

Rawkins, Phillip M. (1978) 'Outsiders as Insiders: The Implications of Minority Nationalism in Scotland and Wales'. *Comparative Politics*, 10(4): 519–534.

Thomas, Alys (1997) 'Language Policy and Nationalism in Wales: A Comparative Analysis'. *Nations and Nationalism*, 3(3): 323–344.

Valenta, Marko (2009) 'Immigrants' Identity Negotiations and Coping with Stigma in Different Relational Frames'. *Symbolic Interaction*, 32(4): 351–371.

Wessendorf, Susanne and Phillimore, Jenny (2019) 'New Migrants' Social Integration, Embedding and Emplacement in Superdiverse Contexts'. *Sociology*, 53(1): 123–138.

Wilson, Elizabeth (1998) 'Planning and Environmentalism in the 1990s'. In Allmendinger, Philip and Thomas, Huw (eds.) *Urban Planning and the British New Right.* London: Routledge. Pp. 21–52.

Winter, Michael (1996) 'Landwise or Land Foolish? Free Conservation Advice for Farmers in the Wider English Countryside'. *Landscape Research*, 21(3): 243–263.

8 *Dych chi'n siarad Cymraeg?* Living with the Welsh language

One of the most significant aspects of moving to Wales was encountering the Welsh language. For many, this was their first proper experience of hearing a second language spoken in everyday life. Although most were aware that Wales had a language, they did not always think about the implications of this. Others did not fully realise that the Welsh language existed at all. The purpose of this chapter is to explore the range of ways these migrants experienced and interacted with Welsh, and how this related to the ongoing social and political developments regarding the Welsh language. Like so much else, Welsh underwent some dramatic shifts in terms of use and status during the 1960s/1970s, and the echoes of these shifts continued to reverberate throughout the following decades. The English, and English migrants to Wales in particular, were embroiled at the centre of these debates and arguments. But, the question of to what extent this actually affected people in their day-to-day lives is one that has received little attention. It is my intention here to continue the story of countercultural migration in Wales by rectifying this oversight.

Before we begin properly telling the story of how this group of migrants coped with the language shift, it is worth taking a moment to think about the words we use to describe these processes. The terms used to ascribe language identities are far from being unproblematic. Bernadette O'Rourke has noted that many of the most common identifiers, including 'native speaker' and 'mother tongue' reflect an inherent assumption of monolinguality. They don't adequately describe identities in multilingual societies, such as Wales (2011: 328). There is a similar level of reductionism in virtually all similar terms: 'learner', 'fluent-speaker', 'semi-speaker', and so on. All of these descriptors oversimplify the reality behind them (Armstrong 2013: 344). 'New speaker' is a relatively recent term that has been coined to help resolve the problems of oversimplification, particularly when dealing with minority languages (Fhlannchadha and Hickey 2018: 39). It is useful here, as it provides a more nuanced way of describing Welsh learners who have reached an operable level of fluency but have limited experience in practice.

Although I have tried to stick to terms like 'new speaker', which negate the inherent assumptions of monolinguality present in dominant terminologies, the attitudes behind languages still remain. O'Rourke et al. have pointed out how 'the history of the native speaker can be traced to anthropologically romantic notions

DOI: 10.4324/9781003358671-12

which link nativeness to a particular community within a particular territory, associated with an historic and an authentic past' (2015: 8). Territory and community are two of the key metaphors used when writing about language, and they highlight the implicit bias which underpins many language terms (Armstrong 2013: 350). No space is left for speakers who originate from different territories, for those who want to join new communities or for territories with more than one 'native' language. This problem does not instantly vanish when the terminologies are changed, and the attitudes associated with language territories and communities are something that we will continue to encounter throughout this chapter.

8.1 Prejudice and decline

When Jim, first introduced in Chapter 2, arrived in Wales, he encountered two initial problems settling in: firstly, he was met with hostility from Welsh speakers due to his lack of language skills, and secondly, there was a growing sense of anti-Englishness in the area he moved to. The idea that local Welsh people were unfriendly towards incomers is one of the most common stereotypes about moving to Wales. The constant underlying tension between England and Wales has led to a belief that all aspects of life in Wales are permeated with this sense of mistrust. Like all stereotypes though, the reality is far more complex. At the first glance, Jim's story seems to perfectly exemplify and reinforce the simplified stereotype version. Jim was a devout churchgoer, both while growing up in London and as a young adult in Kent. On arriving in Wales, he was keen to find a new religious home and saw this as a way of immersing himself into the local community. But when attempting to attend a service, he found himself initially rebuffed due to his lack of language skills:

> When I first came here we went, I attempted to attend a local Welsh-speaking Chapel. So I go to the Chapel, and I can't speak any Welsh at all. . . . I tried several times to learn but erm, I'm afraid it doesn't go with me. But um, I went to a Welsh-speaking Chapel, and I sat through the whole service, which was totally in Welsh which I expected anyway, and a lady came up to me, an old lady came up to me after the service, spoke to me in Welsh, and I said 'I'm terribly sorry, I'm new to the area, haven't learnt Welsh', with which she turned on her heel and walked away.

This story appears to reinforce the notion that Welsh locals were unfriendly towards incomers, but in fact, the cultural issues at play were more nuanced than this. The family had arrived in Wales in 1979, when there were many heated debates over the fate and status of the Welsh language. Today, the Welsh language is a constant presence in Wales, an icon of Welsh culture and identity. This was not always the case. Lesley remembered how when she began taking Welsh classes: 'it was like, 'well, why are you trying to learn Welsh?' because back in those days, there was not the pride in the language that there is now'. Until the late twentieth century, the story of the Welsh language was one of discrimination and decline. The 1536

Act of Union prevented Welsh from being used in courts, as well as preventing anyone using Welsh from holding public office. Because of this, the ability to speak English became associated with importance and high-status – no Welsh speaker could possibly hold as high a position as an English speaker. This bias remained engrained for hundreds of years (Davies 2014: 34–35). Even in the 1970s, there was still concern that using Welsh in schools would lead to poorer language skills (Hornsby and Vigers 2018: 420). The discrimination was therefore still demonstrably alive and well when the countercultural migrants arrived.

All of this led to the number of Welsh speakers declining rapidly. In 1901, 50% of the Welsh population had been able to speak Welsh. Around 15.1% of the population were monolingual Welsh speakers, able to conduct their entire lives through the medium of Welsh with no need for English. By 1981, the proportion of Welsh speakers had dropped down to only 19%, with a mere 0.8% being monolingual (Giggs and Pattie 1992: 269). But Welsh speakers were never spread evenly across Wales. Instead, they were concentrated in certain areas, particularly to the north and west of the country (the Fro Gymraeg of Balsom's 'three Wales' model, which we will return to later on) (Aitchison and Carter 1991: 63–64). These areas were particularly badly hit by rural depopulation and economic decline, identified as two of the major threats facing the Welsh languages (Tunger et al. 2010: 199). As you may remember, these were also some of the key facts which led to countercultural migrants choosing to move to Wales in the first place. This meant that many of the countercultural migrants were in the peculiar position of moving to areas which were some of the strongest Welsh-speaking communities, while also some of the areas most under threat. It should come as no surprise that some locals blamed English immigrants for the loss of the language (ibid). Wales had one of the highest proportions of foreign-born residents out of all constituent parts of the UK in 1981, with 20.6% of the population originating from outside of the country. The equivalent figure for England was 10.2%, Scotland 9.7%, and Northern Ireland 8%. However, in Wales, it was the English who were the largest in-migrant group, making up 17% of the country's total population (Giggs and Pattie 1992: 269).

It was only natural then that some outspoken Welsh speakers should begin to protest at the loss of their culture and heritage. The Welsh Language Movement had been properly kickstarted in 1962, when Saunders Lewis gave his famous speech entitled Y Tynged yr Iaith (The Fate of the Language), which emphasised the ways Welsh had been marginalised by the English. This speech helped to unite Welsh Nationalists and focus their attention on a common goal: the Welsh language (Knowles 1999: 289–290). Thanks to the impact of these campaigners the number of Welsh speakers slowly began to stabilise during the 1980s, and the level of support from the UK government began to increase (Coupland et al. 2006: 352; Giggs and Pattie 1992: 270). Language protection and improved provision for learners formed a key part of this. The 1976 Development of Rural Wales Act made provisions to support a range of Welsh language events and services, including the National Eisteddfod, Welsh Books Council, and *Mudiad Ysgolion Meithrin* (Nursery Schools) (Thomas 1997: 327). Over the last 40 years, Welsh has become more and more strongly engrained in Welsh institutions and law. However, the two most significant events, the passing of the second Welsh Language Act in 1993 and establishment of the Welsh Language Board in 1995, only happened relatively recently (Mann 2007: 213). When the countercultural migrants were arriving in Wales, this was still very much a work in progress. Some encountered ardent nationalists and language campaigners, others like Lesley found Welsh pushed to the sidelines, and others, like Jim, found themselves rejected by Welsh-speaking members of their new community.

Although unpleasant, this unfriendliness caused minimal disruption to Jim's everyday life. He was still able to work and develop his smallholding without interference. Jim's wife, on the other hand, faced the same problem on a more serious level. For her, the inability to speak Welsh had a profound impact on her career and ability to find a job. She had been a primary school teacher when they were living in Kent, and was keen to continue on this career path in Wales. But, for her, the language barrier proved to be a major problem. Jim recalled that:

> She couldn't get any employment here because she didn't speak Welsh, especially in the primary sector. So she, so worked in the local shop from time to time. She did, worked in Rhydygors special school for a while. She was a carer, an invalid carer at that time, but she couldn't work in the education sector because of the Welsh thing, yeah. . . . I think she found it quite hard. Because she's a good, she was a good teacher, and she loved teaching.

Education was a particularly important aspect of the language debates referred to earlier. Immigration formed a crucial part of this, as the arrival of large numbers of predominantly English people led to fears over anglicisation. This, in turn, led to an increased pressure on demands for Welsh-medium education. It was particularly the case in the Welsh speaking heartlands of North and West Wales, but its influence extends elsewhere (Jones and Fowler 2007a: 422). Earlier in the decade the activities of *Cymdeithas Yr Iaith Gymraeg* had focused on introducing bilingual road signs and other similar campaigns, but towards the end of the 1970s, their focus turned to education and other community issues – issues which had a much greater impact on the integration of migrants (Merriman and Jones 2009: 372). This included campaigns against housing and business developments which were judged to be harmful to the Welsh language, casting migrants attracted by such developments in a bad light by association (Gallent et al. 2003: 273).

Jim's wife had problems with employment in Wales because of her lack of Welsh, just as her husband had problems socialising. But language was not the only issue that slowed their integration process. As mentioned earlier, Jim had been a dedicated churchgoer since childhood. The church youth club had formed an important part of his life growing up, and with two young children, the family were now keen to get involved with similar organisations in Wales. However, they found that there was a lack of activities available, and this made integrating into the community especially hard. Jim recalled that, aside from the Chapel, the only options were a football club and history society, both run in Welsh. To try and fill the void,

Jim started a youth club and his wife began a summer playschool. But, although initially successful, this raised its own problems:

> When she did the summer playscheme, which was quite successful, the local authority started to become involved, they said well we'll take it over, put it on a proper footing, so they then advertised that jobs were needed, and we know, we know someone on the interview board, that they gave it to someone who was not a Welsh speaker, but was not English. It meant they'd said, at least she's not English. So she found that quite difficult.

Accounts of such explicit anti-English sentiment as this were relatively rare. But as this shows, it was certainly in existence. It was closely linked to the issues surrounding the Welsh language, both originate from the same fears over anglicisation. There is a strong cultural link between the language and the idea of 'Welshness', and this was a key factor in the resurgence of Welsh nationalism in the 1960s (Williams 1976: 427–428). Anti-Englishness arises, in many cases, from a sense that Welsh identity/culture is being threatened. Since Welsh language is a central part of Welsh identity, and a focus of nationalist campaigns, any perceived threat to it is translated as a threat towards 'Welshness' as a whole. This leads to a sense of anti-English feeling towards incomers who couldn't speak Welsh, even when this was irrelevant to the situation in hand – as was the case with Jim and his wife.

8.2 Welsh classes

All migrants faced a choice as to whether they would try to learn Welsh, and a large proportion of the countercultural migrants decided that they would at least make an attempt. Welsh classes were the most obvious way of doing so. The process of finding and attending a class was not always a smooth one though. For a long time, Welsh as a second language had been taught from a literary perspective, rather than as a functional, everyday language. Few non-Welsh people bothered to try and learn it, as the language had a reputation for being too difficult for second language learners (Newcombe and Newcombe 2001: 332). Some of the earlier migrants to Wales encountered this problem for themselves. Carol, who we met earlier, took Welsh at university in Lampeter:

> It was awful, it was horrible. But it was all a bit literary, there was no . . . it was all stupid things like I could translate seventeenth century poetry but I struggled to order a loaf of bread. I think they've upped their game a bit now, we did used to have language lab lessons and things, but a lot of it was quite correct Welsh rather than the locally used stuff, but it was really difficult. I found it very difficult.

Experiences like Carol's, whether first-hand or heard about through other migrants, could serve to put people off trying to take Welsh classes before they had even begun. During the 1970s, however, the Wlpan method was slowly introduced to Wales. Adapted from the Ulpan system used to teach Hebrew in Israel, this method revitalised Welsh classes with its focus on teaching everyday language intensively, giving the learner the ability to use their skills quickly in real life. Numbers of Welsh learners increased dramatically over the course of the next two decades as a consequence, with many English migrants among the new recruits (Newcombe and Newcombe 2001: 333–336).

However, the process of learning Welsh remained far from easy for everyone. In 1986, Carol Trosset published an ethnographic account of her experience learning Welsh in Wales. She found that locating appropriate classes was still a struggle for many, remarking on how 'most of the people who attempt to learn Welsh (altogether a very small percentage of the English speakers in Wales) must do so without the benefit of more than minimal instruction' (1986: 166). Since many migrants moved to isolated rural communities, the opportunities for learning Welsh could themselves often be severely limited. And even when they were available, they did not always go far enough. Jessica, the vet, found that this was an issue when she was living in Ruthin: 'the problem was that there was, there was a first year and a second year Welsh class, and nothing really after that. After having just done two years you weren't really fluent enough to socialise with Welsh speakers'. It was not until much later, after she moved to Aberystwyth, that Jessica was able to take higher-level Welsh classes. Although some opportunities to learn from home through books, recordings, radio programmes, etc. were available, people tended to dislike them. Judith recalled how

> when I came to Aberystwyth, I thought I'd have a crack at it, but I was only learning through tapes at home, not through speaking to people. And I just, it didn't make sense to me. I couldn't retain what I was reading.

Another issue encountered by migrants trying to take Welsh classes were the tensions between Welsh communities and the (predominantly English) incomers. Language classes have themselves been identified as one of the potential sources of conflict between the two groups, as they automatically separate each community and result in learners being isolated from fluent speakers. This makes it harder for mixed ability relationships to form naturally (O'Rourke 2011: 343). Spending too much time with other migrants, especially those from similar backgrounds to one's own, can result in 'social closure', making it harder to integrate properly into the new community and reducing opportunities to use the new language (Janta et al. 2012: 432). These problems all link back to one of the most significant aspects of learning Welsh: as a minority language, it comes with a different set of challenges to a mainstream language, and the social and cultural dynamic itself is often radically different too (Rampton 1999: 334). It is not as simple as moving to an area where you do not speak the dominant language and must therefore take a class to get a job, with ample opportunities to practise the basics in everyday life. Learning Welsh is very much a choice and doing so requires a tremendous amount of psychological effort.

8.3 Language and integration problems

Under normal circumstances, when immigrating to a country with a different language to one's own, learning the language is a way to help with accessing information and support. Not doing so can result in being excluded from the community (Janta et al. 2012: 431). In Wales, the situation is more complex. Since virtually all Welsh speakers also speak English, not learning Welsh results in far less chance of exclusion (Bermingham and Higham 2018: 399). Indeed, Trosset revealed that finding opportunities to practice was the most difficult aspect of learning Welsh, as Welsh speakers will often switch automatically to English when conversing with anyone who is not fluent:

> [T]he rule most people seem to act on is: When in doubt, assume no knowledge of Welsh, and speak English. Strangers are given one chance to speak Welsh, but if they fail in their attempt they may find it difficult to try again.
>
> (1986: 169–170)

These problems are not unique to Welsh learners, they are shared with learners of other minority languages (Jaffe 2015: 37). Gaelic learners are recorded encountering the same difficulties in trying to practise their language skills as Welsh learners (McEwan-Fujita 2010: 28–30). Some minority languages have coined special terms to describe new speakers, such as Catalan *converts* or Galician *neo-falantes* (O'Rourke and Walsh 2015: 64). New speakers of Breton are described as *neo-bretonnant* while *gaeilgeoir* is a derogatory term for new Irish speakers (O'Rourke et al. 2015: 4). This phenomenon raises two interesting points. Firstly, there does not seem (as far as I can tell) to be an equivalent term for new speakers of Welsh. Trosset suggests that this is because the Welsh language has been maligned for so long that Welsh speakers struggle to comprehend the existence of non-Welsh born Welsh speakers (1986: 172). Secondly, the derogatory nature of the Irish term, combined with the reluctance to speak minority languages with new speakers, suggests some reticence in accepting new speakers into minority language communities.

Stories of the struggle to practise Welsh and be accepted as a Welsh speaker abound in the interview data. Babs, whom we haven't met before, was one of the later migrants to arrive in Wales, moving in 1980. She had previously lived in Zimbabwe (which was Rhodesia at the time) so had plenty of experience settling into an unknown culture, but she found the Welsh language to be a real challenge:

> I attended Welsh evening classes at the primary school, but I only lasted six months! I found it difficult to practise; it seemed that my neighbours felt awkward speaking to me in Welsh plus they said that I was speaking 'posh Welsh' and they didn't use it!.

Her experience was far from unique. Even after the introduction of the Wlpan course, new speakers were frequently criticised for the type of Welsh they spoke.

Corynne, who went on to become a keen champion of the Welsh language, found that when she first moved:

> I tried to learn Welsh, but learnt very old-fashioned Welsh. Um, I went to Welsh classes and it was Welsh that the BBC spoke, and everybody laughed at me. In a nice way, but it still put you off trying to make it, you know.

Corynne may have felt that the laughter was meant in a nice way, but others found it much harder to deal with. Judith remembered how when she arrived in Barmouth:

> I had heard an Englishman being mocked for trying out his Welsh when in a greengrocer's shop. And as he walked away the lady made some snide comments, and it made me feel, 'oh I'm not going to make that mistake'.

Language is something that must be actively 'done'; new linguistic identities need to be actively created and constructed, they do not simply happen (O'Rourke and Walsh 2015: 78). Likewise, acts of 'doing Welshness' form a key part of maintaining a Welsh identity (Coupland et al. 2006: 368). The connection between minority languages and identity were often a key factor in choosing to learn them (Armstrong 2013: 342). But as the aforementioned stories show, existing Welsh-speaking communities often feel they have some control over who counts as a true Welsh-speaker, which can make it difficult for learners to be accepted (Hornsby and Vigers 2018: 423). Learners residing in strongly Welsh-speaking areas face more pressure to learn the language 'correctly' (i.e., according to the standards of the existing community) than those who live in anglicised areas where Welsh is less of a 'community language' (ibid: 425). Trosset notes that the majority of Welsh learners never succeed in gaining more than a basic understanding of the language and suggests that this may be at least partly attributable to the additional psychological pressures that come with trying to learn a minority language (1986: 186–187). Only a small proportion of the contributors to this project reached a stage where they felt confident using the language on an everyday level – most either gave up early on or used their Welsh primarily for a small range of activities, such as participation in Welsh choirs or conversing with bilingual grandchildren.

This links back to the idea that territory and community is significant to language. Fhlannchadha and Hickey have argued that 'questions of authenticity are intertwined with origin, the view that authentic speakers are "from somewhere", and are bound to both geographical roots and a defined community' (2018: 39). But at the same time, many of the gatekeepers to speaking Welsh ultimately came from migrant families themselves, as much of the Welsh population were descendants of those who came to work during the big industrial booms of the nineteenth century (Giggs and Pattie 19992: 269). The line between 'Welsh' and 'not Welsh' is distinctly blurred; however, arbitrary divisions still survive. As discussed in Chapter 2, the way that migrants drew on any kind of familial connection to Wales to justify their decision to move there and ultimately create a Welsh identity for themselves. The same process takes place in trying to validate their position as Welsh speakers. Timothy Armstrong, in his study of Gaelic learners in Scotland, noted that 'interviewees reported that accent, ancestry, and upbringing worked both in their favour and against them at different times and in different situations as they negotiated language use and discursively sought to authenticate themselves as legitimate Gaelic speakers' (Armstrong 2013: 349). Ruby, who was previously mentioned in Chapter 2, was among those who sought to utilise her Welsh connections. Her mother had been born and brought up in North Wales, but left to work as a pharmacist during the Second World War and remained in England afterwards. She had moved back to Wales a couple of years before Ruby herself arrived in the country, and the two lived close together. This automatically gave Ruby some validation in the area, as she recalled: 'It also helped that (as it does in a village) word got round that she was Welsh speaking, even if not strictly local'.

Encountering the Welsh language in everyday use for the first time is one of the few experiences which unite the majority of our group of countercultural migrants. Language traditionally is a key part of migrant experiences, but for this group, it is particularly interesting as they were encountering the Welsh language at a key point in its history. As we have seen, the late twentieth century was the moment when the Welsh language began to revitalise after a long period of decline. Our migrants had to navigate both sides of its history: the prejudice that came with being a Welsh speaker, and assumptions about who can or cannot speak Welsh, as well as the push to start more classes and improve the language's status.

References

Aitchison, John and Carter, Harold (1991) 'Rural Wales and the Welsh Language'. *Rural History*, 2(1): 61–79.

Armstrong, Timothy Currie (2013) '"Why Won't You Speak to Me in Gaelic?" Authenticity, Integration, and the Heritage Language Learning Project'. *Journal of Language, Identity & Education*, 12(5): 340–356.

Balsom, Denis (1985) 'The Three-Wales Model'. In Osmond, John (ed.) *The National Question Again: Welsh Political Identity in the 1980s*. Llandysul: Gomer Press. Pp. 1–17.

Bermingham, Nicola and Higham, Gwennan (2018) 'Immigrants as New Speakers in Galicia and Wales: Issues of Integration, Belonging and Legitimacy'. *Journal of Multilingual and Multicultural Development*, 39(5): 394–406.

Coupland, Nikolas, Bishop, Hywel, Evans, Betsy and Garrett, Peter (2006) 'Imagining Wales and the Welsh Language: Ethnolinguistic Subjectivities and Demographic Flow'. *Journal of Language and Social Psychology*, 25(4): 351–376.

Davies, Janet (2014) *The Welsh Language: A History*. Cardiff: University of Wales Press.

Fhlannchadha, Siobhán Nic and Hickey, Tina M. (2018) 'Minority Language Ownership and Authority: Perspectives of Native Speakers and New Speakers'. *International Journal of Bilingual Education and Bilingualism*, 21(1): 38–53.

Gallent, Nick, Mace, Alan and Tewdwr-Jones, Mark (2003) 'Dispelling a Myth? Second Homes in Rural Wales'. *Area*, 35(3): 271–284.

Giggs, John and Pattie, Charles (1992) 'Croeso i Gymru, Welcome to Wales: But Welcome to Whose Wales?'. *Area*, 24(3): 268–282.

Hornsby, Michael and Vigers, Dick (2018) '"New" Speakers in the Heartlands: Struggles for Speaker Legitimacy in Wales'. *Journal of Multilingual and Multicultural Development*, 39(5): 419–430.

Jaffe, Alexandra (2015) 'Defining the New Speaker: Theoretical Perspectives and Learner Trajectories'. *International Journal of the Sociology of Language*, 231: 21–44.

Janta, Hania, Lugosi, Peter, Brown, Lorraine and Ladkin, Adele (2012) 'Migrant Networks, Language Learning and Tourism Employment'. *Tourism Management*, 33: 431–439.

Jones, Rhys and Fowler, Carwyn (2007a) 'National Élites, National Masses: Oral History and the (Re)production of the Welsh Nation'. *Social & Cultural Geography*, 8(3): 417–432.

Knowles, Anne K. (1999) 'Migration, Nationalism, and the Construction of Welsh Identity'. In Herb, Guntram H. and Kaplan, David H. (eds.) *Nested Identities: Nationalism, Territory, and Scale*. Lanham; Oxford: Rowman and Littlefield Publishers.

Mann, Robert (2007) 'Negotiating the Politics of Language: Language Learning and Civic Identity in Wales'. *Ethnicities*, 7(2): 208–224.

McEwan-Fujita, Emily (2010) 'Ideology, Affect, and Socialization in Language Shift and Revitalization: The Experiences of Adults Learning Gaelic in the Western Isles of Scotland'. *Language in Society*, 39: 27–64.

Merriman, Peter and Jones, Rhys (2009) '"Symbols of Justice": The Welsh Language Society's Campaign for Bilingual Road Signs in Wales, 1967–1980'. *Journal of Historical Geography*, 35: 350–375.

Newcombe, Lynda Pritchard and Newcombe, Robert G. (2001) 'Adult Language Learning: The Effect of Background, Motivation and Practice on Perseverance'. *International Journal of Bilingual Education and Bilingualism*, 4(5): 332–354.

O'Rourke, Bernadette (2011) 'Whose Language Is It? Struggles for Language Ownership in an Irish Language Classroom'. *Journal of Language, Identity & Education*, 10(5): 327–345.

O'Rourke, Bernadette, Pujolar, Joan and Ramallo, Fernando (2015) 'New Speakers of Minority Languages: The Challenging Opportunity – Foreword'. *International Journal of the Sociology of Language*, 231: 1–20.

O'Rourke, Bernadette and Walsh, John (2015) 'New Speakers of Irish: Shifting Boundaries across Time and Space'. *International Journal of the Sociology of Language*, 231: 63–83.

Rampton, Ben (1999) 'Dichotomies, Difference, and Ritual in Second Language Learning and Teaching'. *Applied Linguistics*, 20(3): 316–340.

Thomas, Alys (1997) 'Language Policy and Nationalism in Wales: A Comparative Analysis'. *Nations and Nationalism*, 3(3): 323–344.

Trosset, Carol S. (1986) 'The Social Identity of Welsh Learners'. *Language in Society*, 15(2): 165–191.

Tunger, Verena, Mar-Molinero, Clare, Paffey, Darren, Vigers, Dick and Barłóg, Cecylia (2010) 'Language Policies and "New" Migration into Officially Bilingual Areas'. *Current Issues in Language Planning*, 11(2): 190–205.

Williams, Colin H. (1976) 'Non-Violence and the Development of the Welsh Language Society, 1962–c. 1974'. *Welsh History Review*, 8: 426–455.

9 Identity

Balancing nation and culture

There are two sides to the story of countercultural migrant identity in Wales. Firstly, the question of whether these migrants and their children came to think of themselves as Welsh. Secondly, the question of whether their connection to the countercultural movement had any lasting impact on their identity. Because the move to Wales was intimately connected to their involvement with the counterculture movement, these two strands of their identity are intertwined. Everyone has multiple identities to negotiate – their work persona, home persona, social persona and so on – and these multiple strands of identity can fit together in numerous different ways (Gyberg et al. 2019: 343). This chapter intends to explore the last part of the countercultural migrants' story: the way they continued (and continue) to negotiate their identities and sense of belonging in Wales. Reflection is a key part of identity development; it is how people process their experiences into part of their identity (McLean and Pasupathi 2012: 15; Wilson and Ross 2003). Because of this, the act of being interviewed for this project will itself have impacted on the participants' identity – the way they constructed themselves in interviews may not necessarily equate to how they saw themselves at the time they were speaking about. This is not necessarily a bad thing, but it is worth bearing in mind.

Geographer Mette Strømsø observes that 'the masses of ordinary people in everyday life do not constitute one group of undifferentiated individuals, nor is national belonging evenly distributed within a national space' (2019: 1238). This is certainly the case in Wales. Because people always combined mainstream and alternative cultures, Welsh culture is a central part of the countercultural migrant experience. But, as these stories show, it is not a simple question of 'Welsh' vs. 'English' (or whatever their original nationality may have been). There is a huge amount of variety in forms of 'Welshness', and consequently, there is a huge amount of variety in the ways that countercultural migrants fitted in. Likewise, since not all countercultural migrants possessed the same identity markers, they were not all perceived in the same way. There is a stereotype of English incomers being met by hostile nationalistic Welsh locals, which we touched on earlier, but the stories of this group of migrants suggests that in reality the situation was far more nuanced.

DOI: 10.4324/9781003358671-13

9.1 Belonging vs. identity

Sociologist Floya Anthias has argued that there are several problems with 'identity' as a concept: it struggles to encompass cases of multiple identity and suggests that people who lack a sense of identity, or those whose identity is ambiguous, are problematic (2013: 4). Verkuyten et al., on the other hand, believe that identity suffers from the opposite problem: that it is simply too broad a concept to properly comprehend (2019: 392). Both sets of problems can make it difficult to assess the identity of countercultural migrants. They automatically possess a minimum of two identities (that of their original national identity and that of belonging to the counterculture), and their children often struggled to negotiate new identities for themselves.

Recently, 'belonging' has come to supplement identity as a key concept (Lähdesmäki et al. 2016: 234). The two terms are sometimes used interchangeably, but the distinctions between them are useful here. Geographer Mary Gilmartin argues that 'the concept of belonging offers geographers a way to ground the relationship between migration and identity', and this ability to act as a mediator between these two ideas is particularly valuable here (2008: 1842). Contributors often spoke of feeling at home in Wales, and the connection between 'home' and 'belonging' is naturally very strong (Blunt and Varley 2004: 3). At the same time, these migrants acknowledged that they had no claim on Welsh identity. This absence of identity did not mean that the sense of belonging was necessarily weakened in any way - for many it was extremely strong. Josie expressed it as an almost spiritual connection:

> It's like I know I belong here, it's that belonging to the land, I mean that goes deeper than . . . you know it's a heart thing really. I mean that Welsh word hiraeth? You know that sums it up. I mean I feel free here and I never felt free in south England, I felt enclosed. And there's a feeling I get when I'm coming . . . you know coming over the Severn bridge, and then also turning off the Carmarthen road towards Llandysul, when you're driving into the hills and it just feels . . . it's splendid and glorious.

The feeling of belonging in Wales isn't always an immediate phenomenon for our migrants, but rather something that takes a long time to develop. For Wanda and David (introduced in Chapter 3), the depth of their connection to Wales only really became apparent to them when they moved back to the area after having left to spend some years in the South-West of England:

> It was only when we were driving back behind the um, furniture van coming back from, going from Devon back here, we were driving along the coast and I said 'it feels like we're coming home, it really feels like we're coming home'. It's the most weird feeling but this does feel like home doesn't it?

Barbara was always more aware of the complexity of identity due to both her Northern Irish upbringing and the time she had already spent living abroad. She understood that feeling at home in Wales did not equate to feeling Welsh:

> It is my home now and I feel that. And I'm still Irish, but then I don't, I'm not, I don't bang the drum for nationalism either you know. I think there is a moderation, but I still feel I've my Irish heritage to support.

Acknowledging the role of 'belonging' in countercultural migrants' experiences of living in Wales helps to understand the nuances of the ongoing migrant experience. Mary Gilmartin and Bettina Migge have explained that 'place-belongingness is affective and emotional: it refers to a sense of feeling at home. In contrast, the politics of belonging refers to socio-spatial processes of inclusion and/or exclusion' (2016: 148). Although the role of place-belonging is, as demonstrated earlier, a time-consuming but relatively straightforward process for many, the politics of belonging complicate this. We'll see in the next section how when one begins to take the process of inclusion/exclusion into account it becomes harder and harder to pin-down the identities of this group of migrants.

9.2 Types of Welshness

Unlike legal citizenship, there is no set process by which national identity can be granted. Ross Bond, a sociologist, argues that the three things which are usually regarded as the most important when justifying a claim to national identity or belonging are 'residence, birth, and ancestry', although this varies slightly by nation (2008: 611). Other characteristics that can act as markers of national identity include length of residence, type of upbringing and education, accent, mode of dress, name, and physical appearance (Kiely et al. 2000: 1.4–1.5). Since a large part of the granting of national identity is controlled by those who are already accepted, many of these markers are used (rightly or wrongly) to judge who should be in or out (Bond 2006: 623). In Wales, there is a huge amount of variation in how national identity is expressed and understood. In 1975, Raymond Williams wrote of the confusion surrounding culture and identity in Wales. There were two images, he argued. The traditionalised image of love spoons, *bara brith* and the red dragon, and a second image. This second image was the Wales people actually inhabited, 'a way of life determined by the National Coal Board, the British Steel Corporation, the Milk Marketing Board, the Co-op and Marks and Spencers, the BBC, the Labour Party, the EEC, NATO' (2003: 5).

These divisions had an impact on the sense of national identity developed by our group of migrants; hence, it is worth taking a moment to consider them in more detail. In 1985, Denis Balsom published his now famous (and widely critiqued) 'three Wales model', which tried to map these variations by separating the country into three areas: British Wales, which is the most English of the three, Welsh Wales, and Y Fro Gymraeg, the Welsh heartlands (5–6). Balsom's model is an easy way to quickly illustrate the difference in culture that migrants experienced. For many, it was indeed West Wales which seemed to be the most properly 'Welsh' area (Evans 2007: 135). Heather for instance (who was first mentioned in Chapter 2) moved from Powys (British Wales) to Ceredigion (*Y Fro Gymraeg*) after her children had grown up. When Heather first moved to Wales, she had been reluctant to move this far into the country, 'because before that it was considered very Welsh-speaking, this neck of the woods'. Her experience fits into the model described by Balsom. But, although this may all seem straightforward, the reality is that Wales does not always fit neatly into three categories of identity – and neither do the experiences of the countercultural migrants. Different areas of Wales do indeed each have their own unique identity, and this is often strongly influenced by how much they perceive the area to be anglicised. But there is no clear understanding of what exactly constitutes 'anglicisation', with different people using the concept in a variety of ways, both positive and negative (Evans 2007: 130). This makes the division of Wales into 'Welsh' and 'not Welsh' difficult to achieve in practice. Nobody, including geographers, can agree on an exact boundary delineating the Welsh heartland from the rest of the country (Jones and Fowler 2007b: 94). In practice, the degree of 'Welshness' can change rapidly and fluidly, over relatively short distances. Irene, who we have not met before, arrived in Wales in 1966. After initially living in Newport, she moved to the village of Tregynon: 'it's very English. Very little Welsh there'. But Irene was aware that you did not have to go far for things to change: 'you go five miles to Llanfair Caereinion and it's Welsh'. Five miles really is not far at all, but even this was enough for Irene to identify a difference in culture.

Not only were the boundaries between what was considered 'Welsh' and 'not Welsh' distinctly fuzzy, but people from heavily anglicised areas often still firmly identified as being Welsh in nationality. Being from a 'less Welsh' area did not necessarily mean that someone felt 'less Welsh', and vice versa. There is no clear, universally agreed upon understanding of what being Welsh actually means (Evans 2007: 135; Evans 2019: 178). This could make things very confusing for people moving to Wales. In the latter half of the twentieth century, when our group of migrants were arriving, the outside world struggled to develop an image of a coherent Welsh identity, seeing instead a fragmented picture made up of many different, contrasting elements (Jones 1992: 331–332). This phenomenon is not unique to Wales. Rhys Jones and Carwyn Fowler have argued that 'despite their claims to political, social, cultural, and territorial homogeneity, nations are, and always have been, fundamentally fractured entities' (2007: 351). But it does appear that this phenomenon is particularly apparent in Wales, more so than in other countries, with people struggling to identify a culture which they should fit in with. I observed earlier that many migrants had little idea what they were moving to when they arrived in Wales, and that they often did not realise that Wales would have a unique identity, culture, and language. The difficulty in identifying exactly what Welsh culture was in the first place will not have helped this – and it could make it harder to settle in Wales once the migrants had arrived.

Much of the confusion over what and where 'Welshness' is arises from a tendency to view it as something which is fixed, rather than something which is constantly being made and remade. In looking back to see how migrants fitted in we imagine them, with their own clearly defined identity, embedding themselves in a clearly defined host community. Geographer Anssi Paasi has argued that 'it is common to comprehend regions and localities as *contexts* or *frames* in which social action takes place' (1991: 242, emphasis in original). In other words, geographical research has historically treated the borders of nations as fixed, focusing on the representations of nationhood within them rather than the creation of these representations (Jones and Fowler 2007a: 332). But, just as alternative lifestyles are not clearly defined, so Welsh identity is not clearly defined either.

9.3 National identity in Wales

Despite this, national identity still holds a great deal of significance and value for many. Contributors to the project frequently spoke in terms of whether they identified as Welsh. There is a huge amount of diversity in the contributors' experiences of identity development as a result of the move to Wales. This came about not only because they each experienced a different form of Welshness according to where they settled and who they associated with but also because people can be selective in how they choose to behave after moving across a border. They may be fully comfortable living in a new language but reluctant to participate in politics for instance (Madsen and Van Naerssen 2003: 64). There are those who feel that their country of origin will always dictate their national identity, regardless of how much time they have spent in Wales, or how strong their connection to the country is. Lily, who we have not met before, took a long time to settle into Welsh life:

> It probably took nearly ten years before I truely [sic] felt home going the other way from Bristol to Wales. We now all shout for Wales of course. I would certainly like to be more Welsh than English but I expect I am eternally to [sic] English for that.

Many of those who originally came from other parts of the UK chose to identify themselves as 'British', as this was a way to avoid aligning themselves with either a Welsh identity that they felt they had no right to, or an English/Scottish/Irish identity that they felt they had moved away from. As Lesley, mentioned earlier, recalled:

> I feel British. I don't feel Welsh, England is a bit of an alien country, even now I don't go over, well I go over the border to Shrewsbury, I don't go much further than that. Um, I feel British. I'm not Welsh, I haven't got a drop of Welsh blood, but this has been my home most of my life.

Other migrants had no such qualms about adopting a Welsh identity as their own. For Irene this was one of the things that set her apart from other migrants to the area:

> a lot of the people there had moved in, but they would make no effort at all to learn how to pronounce local names, place names, they'd go back to the Midlands to do their shopping. So, it was just somewhere to live. Well, I wanted to be Welsh.

Corynne was also comfortable enough to take on a Welsh identity: 'I say I am more Welsh than I am English. And I'm proud to be Welsh'. The act of making an active decision on their new place of residence gives migrants an extra sense of belonging, as their new home is part of the life story which they have deliberately crafted for themselves (Antonsich 2010: 649). National identity cannot be granted in the way that citizenship can, if someone wishes to adopt a national identity, they were not born with they must actively do so themselves (Kofman 2005: 454). Many of the contributors who expressed a sense of Welsh identity also, like Irene, expressed a strong desire to become Welsh – it was a key part of the identity they had chosen to develop.

As we know, the contributors to this project came from a range of different countries. Most of them were from England, but there were also migrants from Ireland, Scotland, Zimbabwe (then Rhodesia), Zambia, South Africa, and elsewhere. Migration can sometimes strengthen the links between migrant identities and place rather than disrupting them, and the potential strengthening process tends to play out differently depending on the sense of identity the migrants started out with (Ralph and Staeheli 2011: 521). Before moving to Wales many countercultural migrants will not have had to think much about their own national identity, as they often did not deviate from the standard in their previous homes – it is only when they encountered difference that their own identity becomes apparent (Skey 2011: 233). The exception was those who had either moved multiple times, or who came from countries with stronger or more contested identities. This was the case for Babs and her daughters, who we met earlier. Babs was born in Germany, where her mother was based with the ATS after the Second World War. As an adult she had lived for several years in Rhodesia, before it secured independence and that is where her two daughters were born. The extra cultural shifts gave her a keener awareness of her own place in the world, and she recalled that:

> I probably have a bit of an issue with needing to belong; most likely as a result of my peripatetic childhood; whereas my husband has firm roots in a village in Derbyshire and is sure of his origins. I have recently discovered some Welsh ancestry 3 or 4 generations ago and, at last, I feel I am entitled to live here!

Similarly, the additional movements caused problems for her elder daughter:

> [W]e were never aware of any cultural tension apart from our elder daughter being bullied because she had been 'born in Africa and should be black' (who

knows how she felt; she was and is a sensitive soul whereas her younger sister was a bit tougher).

The upshot of this was that the daughters developed a sense of British, erring toward English, identity, rather than Welsh, despite having spent much of their childhoods outside England. The idea that migrants retain connections to their original culture is not new. This phenomenon has virtually always been recognised by researchers, but the focus has historically tended to be on the connections they form with their new communities, rather than the maintenance (or deterioration) of the old (Vertovec 2001: 574). In some cases, distant existing connections to Wales suddenly took on new importance. People were often quite keen to emphasise these connections, however tenuous they might be, as they helped to ease tensions and give their presence a sense of legitimacy. Laura, who arrived in Wales as a child in 1975, recalled how after moving they discovered that the house they purchased had been owned by a distant relative. This connection, as distant as it may have been, was enough to change the way Laura was received by the community: 'The first day I walked into school we, the dinner lady said to me, 'oh I'm related to you'. And I felt kind of, quite sort of accepted because of that kind of thing, you know'.

Language is one of the most important cultural contributors to developing a sense of belonging. Although it can be used to divide people, it can also help towards developing a sense of community (Antonsich 2010: 648). Language is a key marker of identity, one of the main factors which influences how we are perceived by others, as well as how we perceive ourselves (Feuer 2008: 5). As previously discussed, there was a great amount of variation in how countercultural migrants interacted with the Welsh language. In their interviews, they often implied that there was a link between the ability to speak Welsh and level of Welsh identity. This was overwhelmingly the case when they spoke about their children's identity. When Irene, for instance, was asked if her children considered themselves to be Welsh, she immediately turned to their language abilities:

> Oh yes. Our son . . . they both learnt Welsh at school, [my husband] isn't Welsh speaking because his grandmother was but she didn't speak in Welsh to his mother. They did Welsh at school, our daughter did A-level Welsh, when she first came to Aber she was in Welsh hall of residence.

Corynne also used the Welsh language to back up her children's claims to Welsh identity: 'My children are all Welsh, they've all got their Welsh A-level, all of them, and the youngest got A*'.

It was not just people who had adopted a Welsh identity for themselves that connected their children's ability to speak the language with a claim to identity. Annie, who was mentioned in Chapter 5, makes the same connection: 'My kids, they're both Welsh speakers. One of them, my daughter, moved to Nottingham to go to university and never came home, but she still speaks Welsh and she still considers herself Welsh'. Annie herself was not a Welsh speaker and did not claim a Welsh identity (though she was proud of her connection to the country), but even she recognised the significance of the language as a marker of identity. So did Lily, when speaking about her own children:

> The Eldest lives in Australia. The second lives next door to us with her husband and four children. (She married a Pembrokeshire man). Our third child is passionately Welsh and he is married and lives and works in the Llyn Peninsula [sic] and their children are primarily Welsh speaking. He has learnt good Welsh and works in the medium of Welsh. The last child chose to spend part of his gap year in Bala learning Welsh. He also lives and works in the medium of Welsh in North Wales.

The Welsh language is a crucial part of Welsh identity, so it makes sense that countercultural migrants would draw on it when defining themselves and their families. It provides a sense of connection to the country and its past (Dabrowska 2017: 138). In a sense, it could be argued that the migrants negotiated their relationship with Welsh identity in four ways. Three forms of identity relationship have been discussed already – retaining original national identity, adopting 'British' as an alternative, and claiming 'Welsh' identity outright – and then finally, there was the Welshness which came automatically from being a Welsh speaker.

9.4 (Counter)cultural identity

The question of this group's identity extends further than to what nation they felt they belonged. Identity was a key part of many of the social movements which developed during the 1960s and 1970s (Eschle 2011: 385). Not only that, but they were particularly focused on collective identity – as with many other social movements, their power lay in helping people to develop an identity that united them with others who shared it (Hundeide 2004: 87). The counterculture/alternative movements were no exception. Being part of them, to whatever degree, may have been just as important to these migrants and their sense of identity as their national affiliation. However, contributors to the project spoke much less about identity and belonging in relation to counterculture than they did in relation to nation. Part of this may have been due to me unconsciously swaying the conversation in that direction, but another factor might be that people are less used to thinking of identity in this manner.

One thing that did appear repeatedly throughout the project was participants defining themselves in opposition to other groups – rather than outright statements associating themselves with a particular cultural community, they would refer to other cultural groups that they were definitely not part of. Defining oneself in opposition to 'the other' is far from being a new phenomenon but is often overlooked in research (Triandafyllidou 1998: 593). What is particularly interesting here is that they often defined themselves in opposition to people who were seen to be more countercultural or alternative than they were. Often, this was in the form of references to the mysterious 'hippies in the hills', as Caroline (who was mentioned in

Chapter 6) referred to them. Gay (who was also mentioned in Chapter 6), similarly positioned herself in opposition to this group, recalling that 'It probably helped that I was working in the, the cottage hospital in Tregaron. . . . That kind of made us a little bit more respectable, rather than sort of crazy hippies on the hill'. David, and his wife Wanda, used similar terminology: 'We were never sort of hippies living, living on the hillside, you know, in a tumble-down place'.

Other migrants felt the relationship between these perceived groups was more nuanced. Jo, who we met earlier, recalled how:

> I'm sure I felt in sympathy with a hippy lifestyle, but it wasn't well defined. It really meant you had long hair and wore flowers, and I wasn't, I couldn't have long hair or wear flowers, so . . . . And I didn't drift around in a long dress, 'cause I was very practical, I wore dungarees. So I, I don't think we spent all that much time thinking about where we fitted into a bigger picture, but if we had it would have been more political rather than cultural hippydom.

Uncertainty about exactly what made one part of the hippy-esque counterculture probably partly explains why people were reluctant to identify themselves with it – it was never very clear exactly what they were associating themselves with. Gillian's family reached a novel answer to this conundrum:

> [My son] came home and said 'Are we hippies, mummy?' And so we had this family discussion as to what did we think a hippy was, and were we hippies, and we decided we didn't know. . . . And the first time you have a kid home to tea, and you know we eat in the evening which the Welsh don't, not farmers they have lunch, a dinner. . . . And this child came home and we happened to have a rice dish for tea, our main meal, and the next day he came home and, very happy about it, said 'we are hippies, we eat rice for dinner!' So it's always been a definition of a hippy.

Just as a sense of belonging in Wales often takes a long time to develop, identification with an alternative lifestyle also seems to be something that develops over time. Many people spoke about being identified as hippies now, without the divisions that appear in their accounts of the past. Sue, introduced in Chapter 2, summed this up nicely: 'My youngest daughter goes 'oh god mother, you're such an old hippy', 'yeah I know!", and Corynne made a similar statement: 'An aged hippy, that's what my kids call me'.

The passage of time adds another dimension to migrant identity as it relates to the counterculture. Many of the things which typified countercultural identity when these migrants arrived in Wales – from the differences in diet that Gillian noticed to the reduction in plastic packaging that Jo championed in her wholefoods shop – have gone on to become mainstream. The extent to which the countercultural migrants of the 1970s have engaged with these changes varies greatly by person. Some, like Josie, have refrained from becoming vegan out of respect for the livelihoods of the Welsh community she moved to, reasoning that: 'I'm not going to be completely anti the dairy market because it's a dairy area and people have got to live.' This was an issue that countercultural enthusiasts never had to consider in the days before veganism: do they support the trends they started or the communities where they made their home? Barbara went the other direction, becoming vegan well before it had become trendy: 'I've actually been vegan now for about ten years. Well what's quite shocking now is that I'm almost mainstream, whereas before I was always like . . . it's quite scary being mainstream.'

There are two key processes at play here. Firstly, as people get older, they can find themselves feeling they no longer belong in the place where they are, as it has changed from how it used to be. May describes this process as 'temporal migration', arguing that it can lead to a sense of displacement (2017: 402). The same phenomenon can arise from cultural shifts which occur over time. Secondly, migrants often have a particularly complicated relationship with identity and place. Migration can alter one's sense of identity, but the arrival of immigrants can also alter the identity of the place they move to (Gilmartin and Migge 2015: 83–84). Although the counterculture of the 1960s and 1970s did exist in Wales prior to our group of migrants' arrival their presence, along with that of the many other migrants who were not interviewed here, undoubtedly, had a major impact on the way it was manifested in Wales, and the wider impact the counterculture had on Wales itself (particularly through their entrepreneurial activities). It is also possible that the way the counterculture was perceived by others may have had an impact on people's level of connection to Welsh identity. There has historically been a great deal of concern over a perceived threat to the 'British way of life' posed by countercultural lifestyles (Cresswell 1996: 63–64; see also Halfacree 1996: 45–46; Hetherington 2000; McKay 1996). This means that some migrants may have felt that their lifestyles pushed them *away* from the British identity that they were (often) born into and *towards* Welsh identity.

The identity of countercultural migrants to Wales is a complicated and nuanced topic. There is a huge amount of variation in their experience of identity, both in terms of national identity and cultural identity. For national identity, these ranged from a strong connection with their place of origin to complete adoption of a Welsh identity. But this simple scale of 'levels of Welshness' is complicated by several factors. Some felt an extremely strong sense of belonging in Wales, without feeling a sense of Welsh identity. Others ascribed Welsh identity to their children based on ability to speak Welsh, creating a strong link between the two.

When it comes to identity and counterculture there is a similar phenomenon, but it is still less clearly defined. People defined themselves in opposition to others more clearly than in the case of national identity and expressed a greater sense of uncertainty about exactly what constituted the counterculture. Because of this, not everyone who identified with it had the same understanding of what they were part of – although this did not necessarily diminish the strength of their identity. And just as their sense of belonging to Wales tended to strengthen over time, so too did people's association with the counterculture. This was partly due to the things which were considered countercultural in the 1960s and 1970s becoming more mainstream over time. In a sense, identity provides a microcosm of the countercultural migrant experience. It encompasses each of the key elements: from the reasons they chose to move (to become Welsh, to be more countercultural, and so on), to the specific way they engaged with the counterculture, and to the relationships they formed in their new Welsh communities. It shows the breadth and variety of these experiences, and the way these impacted on the migrants themselves.

The stories shared in these three closing chapters exemplify the diverse range of integration experiences. A number of factors impacted on all countercultural migrants to Wales, such as the stereotypes surrounding the Welsh language, and complexities of the 'local' and 'migrant' identity categories. I have explored how the interaction between migrants and these factors reflects different theories around migrant interaction. We have seen how exchanges of social capital were particularly significant to countercultural migrants to Wales, but in a different way to the exchanges of social capital elsewhere. Knowledge sharing between locals and migrants seems to have been one of the most successful integration tactics. Providing needed services, like Barbara did, was also valuable. But the tensions surrounding incomers and the fear of anglicisation complicated this; as we saw in the case of Jim and his wife, providing a service was not a guarantee of acceptance. Children were another key part of early integration experiences, but despite their usefulness to parents they often struggled with the process themselves. Their stories in particular highlight the othering of migrants by locals, an experience which often affected adults less severely. The Welsh language could both act as an aid to integration, and make it harder to socialise. These experiences all combined to create the diverse range of identities expressed by the countercultural migrants.

These three chapters have shown that countercultural migrants to Wales faced a number of challenges when trying to fit into Welsh communities. Living in Wales meant negotiating a complex set of socio-cultural expectations and assumptions, with a long and fraught history. The migrants' ability to fit in was determined as much by the history of the places they moved to as it was by their own behaviour or the way they were perceived by the locals. Everything, from the level of 'Welshness' they encountered (if there is even such a thing), to the assumptions made about their ability (or inability) to speak Welsh, was impacted by external and internal factors. It is no surprise, therefore, that so many people felt they needed some justification to be there, some connection that would allow them (or their children or grandchildren) to feel that they belonged. At the same time, it cannot be said that the behaviour of the migrants did not also have an impact. They had to actively choose how they were going to fit into these communities, what role and identity they were going to take up. Decisions about whether to learn Welsh, to mix with locals or migrants, or to define themselves as Welsh still fell, ultimately, to the migrants themselves. And through this the diversity of migrant experiences and aspirations becomes apparent. Each one has an entirely unique story, which shines a light on different aspects of Wales and the counterculture.

References

Anthias, Floya (2013) 'Identity and Belonging: Conceptualisations and Political Framings'. KLA Working Paper Series, No. 8.
Antonsich, Marco (2010) 'Searching for Belonging: An Analytical Framework'. *Geography Compass*, 4(6): 644–659.
Balsom, Denis (1985) 'The Three-Wales Model'. In Osmond, John (ed.) *The National Question Again: Welsh Political Identity in the 1980s*. Llandysul: Gomer Press. Pp. 1–17.
Blunt, Alison and Varley, Ann (2004) 'Introduction: Geographies of Home'. *Cultural Geographies*, 11: 3–6.
Bond, Ross (2006) 'Belonging and Becoming: National Identity and Exclusion'. *Sociology*, 40(4): 609–626.
Cresswell, T. (1996) *In Place/Out of Place: Geography, Ideology, and Transgression.* Minneapolis; London: Minnesota University Press.
Dabrowska, Izabela (2017) 'The Role of Cymraeg in Shaping Welsh Identity'. *Anglica: An International Journal of English Studies*, 26(1): 131–147.
Eschle, Catherine (2011) '(Anti-)Globalisation and Resistance Identities'. In Eilliott, Anthony (ed.) *Routledge Handbook of Identity Studies*. London: Routledge. Pp. 291–307.
Evans, Dafydd (2007) '"How Far across the Border Do You Have to Be, to Be Considered Welsh?": National Identification at a Regional Level'. *Contemporary Wales*, 20(1): 123–143.
Evans, Daniel John (2019) 'Welshness in "British Wales": Negotiating National Identity at the Margins'. *Nations and Nationalism*, 25(1): 167–190.
Feuer, Avital (2008) *Who Does This Language Belong To? Personal Narratives of Language Claim and Identity*. Charlotte, NC: Information Age Publishing.
Gilmartin, Mary (2008) 'Migration, Identity and Belonging'. *Geography Compass*, 2(6): 1837–1852.
Gilmartin, Mary and Migge, Bettina (2015) 'Home Stories: Immigrant Narratives of Place and Identity in Contemporary Ireland'. *Journal of Cultural Geography*, 32(1): 83–101.
Gilmartin, Mary and Migge, Bettina (2016) 'Migrant Mothers and the Geographies of Belonging'. *Gender, Place and Culture*, 23(2): 147–161.
Gyberg, Fanny, Frisén, Ann and Syed, Moin (2019) '"Being Stuck between Two Worlds": Identity Configurations of Occupational and Family Identities'. *Identity*, 19(4): 330–346.
Halfacree, Keith (1996) 'Out of Place in the Country: Travellers and the "Rural Idyll"'. *Antipode*, 28(1): 42–72.
Hetherington, Kevin (2000) *New Age Travellers: Vanloads of Uproarious Humanity*. London; New York: Cassell.
Hundeide, Karsten (2004) 'A New Identity, a New Lifestyle'. In Perret-Clermont, Anne-Nelly, Pontecorvo, Clotilde, Resnick, Lauren B., Zittoun, Tania and Burge, Barbara (eds.) *Joining Society: Social Interaction and Learning in Adolescence and Youth.* Cambridge: Cambridge University Press.
Jones, R. Merfyn (1992) 'Beyond Identity? The Reconstruction of the Welsh'. *Journal of British Studies*, 31(4): 330–357.
Jones, Rhys and Fowler, Carwyn (2007a) 'National élites, national masses: oral history and the (re)production of the Welsh nation'. *Social & Cultural Geography*, 8(3): 417–432.
Jones, Rhys and Fowler, Carwyn (2007b) 'Where Is Wales? Narrating the Territories and Borders of the Welsh Linguistic Nation'. *Regional Studies*, 41(1): 89–101.
Kiely, Richard, McCrone, David, Bechhofer, Frank and Stewart, Robert (2000) 'Debatable Land: National and Local Identity in a Border Town'. *Sociological Research Online*, 5(2).
Kofman, Eleonore (2005) 'Citizenship, Migration and the Reassertion of National Identity'. *Citizenship Studies*, 9(5): 453–467.

Lähdesmäki, Tuuli, Saresma, Tuija, Hiltunen, Kaisa, Jäntti, Saara, Sääskilahti, Nina, Vallius, Antti and Ahvenjärvi, Kaisa (2016) 'Fluidity and Flexibility of "Belonging": Uses of the Concept in Contemporary Research'. *Acta Sociologica*, 59(3): 233–247.

Madsen, Kenneth D. and Van Naerssen, Ton (2003) 'Migration, Identity, and Belonging'. *Journal of Borderlands Studies*, 18(1): 61–75.

McKay, George (1996) *Senseless Acts of Beauty: Cultures of Resistance since the Sixties*. London: Verso.

McLean, Kate C. and Pasupathi, Monisha (2012) 'Process of Identity Development: Where I Am and How I Got There'. *Identity*, 12(1): 8–28.

Paasi, Anssi (1991) 'Deconstructing Regions: Notes on the Scales of Spatial Life'. *Environment and Planning A*, 23: 239–256.

Ralph, David and Staeheli, Lynn A. (2011) 'Home and Migration: Mobilities, Belongings and Identities'. *Geography Compass*, 5(7): 517–530.

Skey, Michael (2011) '"Thank God, I'm Back!": (Re)defining the Nation as a Homely Place in Relation to Journeys Abroad'. *Journal of Cultural Geography*, 28(2): 233–252.

Strømsø, Mette (2019) 'Investigating Everyday Acts of Contributing as "Admission Tickets" to Belong to the Nation in Norway'. *Nations and Nationalism*, 25(4): 1238–1258.

Triandafyllidou, Anna (1998) 'National Identity and the "Other"'. *Ethnic and Racial Studies*, 21(4): 593–612.

Verkuyten, Maykel, Wiley, Shaun, Deaux, Kay and Fleischmann, Fenella (2019) 'To Be Both (and More): Immigration and Identity Multiplicity'. *Journal of Social Issues*, 75(2): 390–413.

Vertovec, Steven (2001) 'Transnationalism and Identity'. *Journal of Ethnic and Migration Studies*, 27(4): 573–582.

Williams, Raymond (2003 [1975]) 'Welsh Culture'. In Williams, Daniel (ed.) *Who Speaks for Wales? Nation, Culture, Identity*. Cardiff: University of Wales Press. Pp. 5–11.

Wilson, Anne and Ross, Michael (2003) 'The Identity Function of Autobiographical Memory: Time Is on Our Side'. *Memory*, 11(2): 137–149.

Conclusion

The aim of this book has been to answer a set of research questions that fit into three broad themes. Firstly, motivations: why did so many people move to rural Wales during the late 1960s and 1970s? What drove them to leave their previous homes, and what made them settle in Wales specifically? Secondly, practices: what kind of alternative knowledges and beliefs did they bring to rural Wales, and what was it like putting these into practice on an everyday level? Finally, belonging: how did these migrants fit in with existing communities? What impact did living in rural Wales have on their identities, and those of their descendants? In answering these questions, I have tried to make sure the focus always stays on the stories shared by the contributors, linking back to them whenever I have paused for a contextual or theoretical discussion. The focus on individual lives and lived experiences is a core part of this project, intrinsic to its methodology and rationale. It means that the answers to these questions, stemming from the individual accounts of each contributor, are far from clear-cut. However, it also means that it is easier to see the full picture of how each aspect of their experience – encompassed by the research questions – connects with each other.

The reasons why so many people were keen to move to the countryside can be broadly summarised into 'push' and 'pull' factors to use the language of classic migration studies. This binary has been widely discarded as too simplistic, but it provides a useful way of summarising the content of this book. The stories shared here have shown that both sets of factors were strongly linked to the social and political events of the 1960s and 1970s. These stories highlighted a discomfort with certain aspects of contemporary life, especially for those living in urban or sub-urban environments before their move – these were the push factors. There were fears over safety, prompted by IRA attacks, frequent and disruptive strike action, and so on. There was also the pervading sense of discomfort with urban culture at the time, a feeling of not fitting in, a disinterest in rising consumerism and its increased social significance, and upset at what was seen to be a declining sense of community. Frustration at the political and economic chaos that accompanied these changes led people to try and imagine a new way of life, away from a political system that felt like it was failing, and away from urban environments that were seen as unsafe and unfriendly.

DOI: 10.4324/9781003358671-14

On the other hand, were the things that made life in rural Wales seem like an appealing alternative – the pull factors. The countryside offered the potential for buying the land needed to practise a different way of life, incorporating elements of self-sufficiency. The stories also recall a sense of being attracted to the greenery and beauty of rural landscapes – and to the sense of wildness, desolation, and emptiness that was especially felt in Wales. There was an idea that life in the countryside was somehow intrinsically better than anywhere else, a lingering nostalgia for 'simple' rural communities as described by Tönnies (2012 [1887]: 16). Although these factors may seem entirely separate from the political and social factors that were described earlier, the stories have shown that there was an intrinsic connection between the two. Since rural Wales was experiencing drastic depopulation at this time there was an increase in empty houses being left behind – both providing attractively inexpensive housing options and giving rural Wales its characteristic sense of being wild and empty. Then, there were also the initiatives to develop rural Wales in the wake of its depopulation. This was one of the more unexpected factors uncovered – many migrants were attracted to Wales not so much because of its potential for alternative lifestyles but also because of the programmes designed to encourage investment in and migration to the area.

There was of course one final factor, which does not clearly fit into 'push' or 'pull' categories, but is important to return to because of its implications: the fact that DIY, and by extension the ideas of self-sufficiency enthusiasts like John Seymour, were very much in vogue at this point – even outside of the counterculture. The stories shared here highlight how many of the migrants ended up buying badly run-down or derelict houses, with the intention of doing them up themselves. This was part of a wider trend, as rising house prices led to more and more people choosing to improve their homes rather than move or pay someone else to do it. What this shows, along with the other connections between the decision to move and the wider events of the time, is that the line between 'mainstream' and 'counterculture' is distinctly blurred. The two do not exist as clearly definable separate entities but overlap and intersect in unexpected ways. Moving to rural Wales may have been considered an 'alternative' thing to do, but among the contributors to this project, the decision to move was driven by wider social norms just as much as it was driven by a desire to break those norms. In the literature review, I noted that Keith Halfacree has observed a gap in migration literature concerning the counterculture (2009: 771). Through answering the question of why so many people moved to rural Wales during the late 1960s and 1970s, this project has not only helped to fill this gap but has also helped to shed new light on the decision-making process that lies behind wider lifestyle and counterurban migration movements.

The answers to the second research theme's questions – what kind of alternative knowledges and beliefs did this group of migrants bring to rural Wales, and what was it like putting them into practice on an everyday level? – are connected to the answers to the first theme's questions; both of them share an unexpected blurring of 'alternative' and 'mainstream'. The discomfort with contemporary life that prompted so many decisions to move, that pervading sense of not fitting in, was also what prompted the development of new lifestyles that would provide an alternative – this was, after all, precisely what the counterculture was all about. In reacting against consumerism migrants increased their reliance on home production and traditional craft skills, drawing on a nostalgic 'make do and mend' ethos. Interest in home production led some of them to aspire towards self-sufficient lifestyles, inspired by the work of John Seymour and other authors such as Allaby and Tudge (1977), Cherrington (1983), or West (1977) (and tying back to the growing 'mainstream' interest in DIY discussed in Chapter 3). This in turn led to a renewed emphasis not just on crafts but also on the weather and the seasons – the changing patterns of nature that are so important to living off the land but were rapidly becoming replaced by more important concerns in other people's lives. This part of the stories paints a gentle, nostalgic image – a seemingly perfect life of reconnecting with animals and nature, rediscovering lost crafts, of feeling that whilst money may be in short supply, time was most certainly not. However, of course, this is not the whole story – not for even those most committed to this branch of countercultural lifestyle. The stories shared here have highlighted how this was especially evident in the role played by women in the counterculture, who found themselves dealing with a conflicting set of expectations around the fiercely gendered role of rural women on one hand and the feminist branch of the counterculture, exemplified through publications like *Spare Rib*, on the other.

We must not forget though that practising self-sufficiency and going 'back-to-the-land' was not the only 'alternative' lifestyle option, and only in rare cases did it make enough to live on; hence, the migrants' choice of employment was another big theme in their stories. Many individuals took on jobs that would be considered 'mainstream', again demonstrating just how blurred the boundary with the counterculture really was. A considerable number started their own businesses in Wales. The connection between entrepreneurship and the counterculture may not seem obvious to us today, but these stories have highlighted how the differences in public opinion around entrepreneurs in the 1970s made starting a business a far more rebellious act than it is now. This fact sheds light on an important aspect of these migrants' stories – because the 'mainstream' culture is always changing, things which have been considered 'countercultural' at one point in time may not always continue to be so (and vice versa). Diet was a big part of the knowledges and practices that these migrants brought to Wales. Many of them recalled how they were unable to source the international foods they had previously enjoyed eating, and had to resort to selling or growing them themselves – thus also making them available to the wider Welsh community. Since many followed vegetarian or macrobiotic diets, businesses catering to these choices also sprung up where the migrants arrived. Just as with entrepreneurship in general, these are foods that are not regarded as especially unusual or exotic today, but they certainly were then.

There are other threads to this part of their story, other aspects to the practicalities of living 'alternative' lives. There is the role played by religion, the tendency of these migrants to be drawn to Buddhist or Quaker beliefs, the challenge of balancing one's own faith with a desire to respect the beliefs of local communities. There was the role played by the underground press, the want to engage with this classic part of the counterculture problematised by its urban focus. Tying together all of these threads are the two findings I have highlighted – that the line between 'counterculture' and 'mainstream' is blurred, and that the very definition of 'counterculture' is entirely dependent on context.

The migrants' stories become even more complex when we turn to the answers to the final question: how did they fit in with existing communities, and what impact did living in rural Wales have on their identities? These stories have revealed many different, often conflicting, answers. On one hand, the arrival of the migrants could be quite positive, and there are plenty of stories of them being welcomed by local communities. Because many individuals came as part of attempts to revitalise depopulated areas, and because they made permanent moves rather than being second-home owners, they recalled feeling more committed to the move and thus more connected to the existing communities. In some ways, this links back to the second question, as those who started businesses or were involved in introducing new foods found that the extra investment in developing the local economy also helped them fit in.

One of the most memorable insights to come out of the interviews was Barbara's observation about fitting into the 'spaces in between'. The migrants' economic integration to the area can be seen as an example of this, as they actively engaged with their communities and took the initiative to provide services where there were gaps. However, the idea of filling the 'spaces in between' goes further than economics; it also extends to the way these migrants behaved within their communities, and the social activities they undertook. They often shared memories of participating in organised forms of socialisation, and frequently participated in social groups – or started their own. Their arrival raised the population levels enough to keep many otherwise-threatened rural schools open, and they became involved with school activities, as well as facilities for children below school-age. These are all ways that migrants used their lifestyle choices to fit in with existing communities rather than imposing themselves on them – ways of fitting into the 'spaces in between'.

Then, of course, there were the less positive stories of trying and failing to fit in. We have already touched on the tensions which could arise over the role of women in a staunchly masculine rural setting. Those who moved as children in particular had a difficult time of it. With less freedom to choose how, and how much, they interacted with their new communities, and more pressure to fit in with pre-existing social groups in school, children faced unique challenges. Perhaps some of the biggest tensions, certainly those most widely discussed, arose around the Welsh language. The stories show a mixture of responses to encountering the Welsh language, with some embracing it wholeheartedly (more so than Welsh-born residents in some cases), some finding themselves held back by their inability to grasp the language, and others merely ambivalent. Their experiences highlighted some of the wider issues surrounding attitudes towards the Welsh language, with many of those who did try to learn recalling either that they struggled to practise because people were unused to speaking Welsh with new residents, that when they did practise they found the variety of Welsh they had been taught did not match the local dialect, or that their new neighbours were not Welsh speakers themselves anyway.

In many cases, as noted earlier, the migrants' decision to move to Wales was driven in part by an idealised and unrealistically positive image of rural communities. Their experiences in reality reinforce what was evidenced by the literature review: that community can actually be a negative, exclusionary concept (Smith 1999; Day 1998). However, they also show that there was far less negativity around than might have been expected from earlier literature, especially when one considers the well-recorded tensions between Wales and England at this time. Although there were certainly ups and downs in the migrants' interactions within their new communities, their stories suggest that, for the most part, people were prepared to set aside their differences on a day-to-day level.

The way they negotiated their relationship with the Welsh language impacted not only the migrants' relationship with their new communities but also the identities they defined for themselves. We have seen that the impact of moving to Wales on their identities was just as diverse as everything else, but there is a clear pattern of associating an ability to speak Welsh with adopting a Welsh identity. Although the migrants' identities shift and change over the years, as the migrants become more or less attached to different places and cultures, the stories do not suggest that the migrants respond to the changes in society in the same way that the wider counterculture does. In contrast, the extent to which the migrants identify as 'countercultural' – a term I will be unpacking more fully later on – is very much dependent on external factors. We have seen how many stories make the observation that their 'alternative' lifestyles - the things which make them 'hippies' as many of them put it – have become more and more mainstream. As this happens, the migrants must deal with the subsequent shift in the way their identities are perceived. The observation that identities are continually made and remade is not new – but the stories shown here give a striking insight into the everyday experience of these processes, and reinforces the significance of the combination of time and place to the development of identity.

Like any research project, this one has had its limitations. It has been impossible to cover everything that I would like to in the time and space available. Although I have done my best to tell as complete a story as possible, there remain ways that it could have been expanded. I spoke with 55 individuals during the interview process, and this was more than enough to provide the amount of data needed. A large number of stories were only able to influence the narrative and could not be told individually, which is a great shame. We will have to wait for another chance to hear about the man who came to Wales after being on a council with Nelson Mandela, we would need far more space to go into detail about another migrant's run in with an Afghan assassin, and those are just two examples out of dozens. Telling them all would have been deeply impractical; hence, I have stuck to the ones which seemed to best answer the research questions – although doing my best to make sure that as many of the interviewees as possible got their voices heard somewhere. On the flip side, there is always the possibility that more people could have been interviewed, to get a broader range of experiences. This might well have revealed different information, highlighting different commonalities between the migrants' experiences. There is no way of knowing just how representative this group of migrants were. However, since my goal was not to develop a generalisable image, I do not consider this to be too much of an issue.

In the introduction, I pointed out the similarity between the Black Lives Matter movement, which is going on as I write, and the Civil Rights Movement that was happening at the time the first countercultural migrants were moving to Wales. You may have noticed that race hasn't been mentioned much since then. The group of migrants I interviewed were all white, and the implication from their stories is that the communities they moved to were also predominantly white (an implication which is corroborated by the general demographics of Wales at this time (Robinson 1993: 150)). This leaves us with two possible questions unanswered. First, were there any BAME migrants who were simply not picked up by my push to find interviewees? It is entirely possible that there were, and this is something that it would be well worth investigating further (although it could turn out that there were not). Second, what was the relationship between BAME communities and the counterculture, or more specifically, between BAME communities and this particular group of migrants? The interview data gives hints as to the answers to these questions, but there is nothing conclusive. One interviewee's story of deciding to leave their home in the town suggests that there were undertones of racism in the decision to move to the countryside:

> Um, what I didn't like when living in [the town], I knew the people either side of me but I didn't like them very much, um, I knew nobody else in our street. And [the town] was becoming a very coloured area. I've got nothing against coloured people, but they tend to be in groups, you know, and so they would exclude whites. You know, and they waged, they made the colour divide. They made it, you know. And then of course you come out to Wales where there was nobody.

It would not be entirely surprising if such opinions were shared more widely among countercultural migrants to Wales, given that the Welsh countryside continues to be perceived as a predominantly white area 'free' from issues of race (for more discussion of the problem of rural racism, see Williams 2007 and Chakraborti 2010). This is a difficult topic to broach in interviews, and one which requires a great deal of forethought and sensitive handling. People are hardly going to admit to being racists outright, and fairly unlikely to suggest that their friends or neighbours are either (although one contributor was quite happy to inform me that on arriving in Wales they discovered that the local headmaster and his children were all racists). There is evidently more to be uncovered about the role of race in the counterculture, and it will take a second research project to do justice to the full story.

This research has been very much rooted in Wales, with the depopulation of rural Wales and subsequent attempts to revitalise the area forming a key part of the backdrop to these migrants' experiences. However, countercultural migration was certainly not limited to Wales. Similar movements occurred in other rural areas of Britain and Ireland, as well as internationally. These will likely have encompassed many of the same themes, but there will be key differences around the specifics of each case (especially relating to identity and language). As with so many other things, including a wider geographical area would have pushed this project beyond its limits – but it could help pave the way for future research. Likewise, this project has reinforced the significance of international influence on the countercultural movement – and on the migrants to Wales specifically – especially the impact of going on the hippy trail. The importance of international influences has cropped up many times throughout these stories. There was the significance of macrobiotics, and heavy reliance on imported or relatively unknown ingredients in the countercultural diet, which led to the introduction of new foods to Wales. Then there were the other import/export businesses that were set up, bringing goods from Afghanistan and elsewhere. Buddhism, too, was an international influence on counterculture. Let us not forget that not all of the countercultural migrants came from the UK in the first place – this was an international migration as well as an internal one. I'm sure there are more stories to be uncovered here. The hippy trail and its wider impact are something that has received precious little attention in existing literature (with the exception of Gemie and Ireland's (2017) book on the topic), but it clearly had great significance to many people. This is another way that the geographical reach could be expanded, building on the principles and findings of this project.

Finally, perhaps the most obvious question that is raised by this research, with its emphasis on the experiences of individuals at ground level, is what were the pre-existing Welsh communities' experiences of the countercultural lifestyle migrants' arrival? We have caught glimpses of this side of the story through the migrants' accounts of settling into their new communities and through the wider history of Welsh language protests and related movements. However, there is potential for a parallel study of the lived experiences of host community members, collecting their stories and seeing how they merge with, or diverge from, those told here.

Hence, what are the key findings of this research? What is the overarching conclusion that you can take away from it? One of the clearest findings of this project is that the very idea of a 'counterculture' is something that is entirely dependent on context. I began writing the original version of this conclusion on the night of the 2020 US Presidential Election. By now this is old news, but at the time, there was a pervading sense of tension everywhere. The fate of the world was tossed into the air, and we could only hold our breath and wait to find out what would happen when it fell back to Earth. One cannot help but wonder how many times the people who moved to Wales found themselves anxiously watching the world like this, witnesses to the turning points in time, knowing their lives were being buffeted by intangible forces beyond their control. We know now, having listened to and analysed their stories, that for many migrants a sense of dissatisfaction, frustration, sometimes even fear, about the way the world was going was a key factor in the decision to move to Wales. It has what the counterculture was based on after all, the idea of finding an alternative. Perhaps, they sat and watched news reports of the IRA attacks that would lead them to think of cities as unsafe homes or anxiously awaited to hear the results of the Government's negotiations with trade unions during the many strikes that threatened the stability of the country. They almost certainly felt the tension of election night on plenty of occasions, as political unrest meant there were far more elections between 1965 and 1980 than ought to have been the case.

Crucially though, this group of people didn't just sit still while time happened around them – they were actively living their lives in response to these events. Maybe some events had more impact than others, and different people will naturally have been affected by different things, but nobody was just a passive observer. Perhaps, at its heart, this is a story of what happens to people as the world changes, how individuals cope with, and respond to, the shifts occurring around them. I may have expected to find a consistently identifiable movement, but this was far from the case. For Roszak, in his 1969 book on counterculture, it was technology and technocracy that was the 'enemy' to be rebelled against. But in stories shared here, most of which occurred just a few years later, it was capitalism that was the most obvious adversary. Writing 14 years after Roszak, Yinger abandoned the idea that counterculture could be defined as opposition to any single specific thing, instead using the term to refer to 'a set of norms and values of a group that sharply contradict the dominant norms and of the society of which that group is a part' (1982: 2–3). Like Yinger, I chose to define 'counterculture' as simply a resistance to the 'mainstream', rather than linked to any specific practices or beliefs. I initially did this because I wanted to leave the term open, to allow space for people to fill with their own stories. They certainly did. The counterculture ended up being far broader even than I had expected, a weaving together of a multiplicity of disparate threads. I had not expected it to be something that would change so much according to time and place. It is obvious when you think about it; if counterculture is opposition to the mainstream, then naturally it would change according to the mainstream culture/s of the time.

Because perceptions of what is 'mainstream' is linked to place as well as time, there is an especially complex relationship between the countercultural migrants and the places they moved to. Things that may have been countercultural in their original homes – such keeping livestock – may be perfectly normal in the rural settings they moved to. Likewise, things which were not so countercultural before the move – eating foreign foods for instance – could suddenly become an unintentional act of rebellion. This is not just a culture shock, but a balancing of two different sets of 'mainstreams' and 'alternatives'. The wider counterculture had a very urban focus, as we know from the earlier discussion of the underground press (Nelson 1989). From the stories shared here it seems that, although the counterculture did shift in relation to temporal variations in the mainstream, it did not account for spatial variations. Instead, the stories shared here suggest that these variations were utilised as a way of developing the idealised 'alternative' lifestyle the counterculture endeavoured to create. Rural life was seen through the lens of the idealised image of the countryside, and this was something the counterculture tried to co-opt as a rejection of the 'mainstream' that was present in the (predominantly urban and suburban) areas where the counterculture was at its strongest and most well-known. I do not think the migrants consciously thought through the process like this – indeed, I doubt many people involved with the counterculture ever put much thought into the division between 'mainstream' and 'alternative'.

The concern was on finding solutions to what were seen as problems with the lifestyles that they were living – it just so happens that this is how they ended up in practice. The implications of this on a personal level are rarely considered – it means that if an individual wishes to consistently participate in the counterculture they must constantly shift their lifestyle accordingly. As the migrants in the final chapter noted, maintaining the same lifestyle over the years since their move means that, as culture has changed around them, they have now found themselves aligning with the mainstream.

As discussed earlier, spatial differences in social norms were often utilised by the counterculture rather than rejected as part of the maligned 'mainstream'. There is something almost colonial in the way the counterculture borrowed indiscriminately from others, picking and choosing what to reject and what to embrace. This project has shown that not only was the idea of 'counterculture' entirely dependent on context but also was specifically dependent on the temporal context. When it came to cultural differences within this, it became a colourful conglomeration of different ideas – not consciously selected by any one person, or group of people, but brought together by the idea that they somehow aided in the quest to find an 'alternative' to a 'mainstream' that was understood in predominantly urban Western terms.

Of course, not all of the cultural differences were known or understood by countercultural migrants, and they experienced complex processes of trying to fit into their new communities. The legacies of this generation of in-migrants trying to fit into previously unknown Welsh communities, with their distinctive Welsh cultures and ways of life, continue to be felt in the present. In-migrants have continued to arrive, and they have faced increasingly complex histories of relationships between migrants and local communities. The counterculture plays a key part in this, and in particular the history of trying to reconcile one dominant construction of rurality with that present in rural Welsh communities (Cloke et al. 1998: 140–144). The countercultural migrants to rural Wales left other legacies too. This was not a one-time event, it is an ongoing process, part of the creation of the Wales we know today, and influencing the Wales that we will see in the future. There are the obvious legacies of course, the people themselves, their descendants, the businesses which have thrived and survived into the present. Less obvious legacies exist as well, like that of constructions of rurality. There is the lingering impact of countercultural influence on the culture of rural Wales, and the way it is perceived by others. For better or worse, the idea of Wales as a hotspot of 'alternative' lifestyles has thoroughly permeated people's imaginations. Keith Halfacree describes how, as a result of in-migration to the area, 'rural Wales contains a thriving set of geographies of "alternative" lifestyles or alterity. This sense of otherness *is* acknowledged by popular culture, with "hippies" now a key element within representations of rural Wales' (2011: 67–68, emphasis in original). The countercultural migrants' legacy in rural Wales is not straightforward or easily definable, any more than the act of countercultural migration was itself straightforward or easily definable.

This leads us to the second key finding from this research: the line between 'mainstream' and 'counterculture' is inseparably blurred. The fact that so much of rural culture was appropriated as part of the counterculture is a key example of this. Welsh culture and language were clearly a significant part of these migrants' experiences, but they had no real connection to the counterculture as it is traditionally understood. It highlights the blurring of the boundaries between 'mainstream' and 'counterculture' on two levels. On the one hand is the messiness of ordinary lives – the fact that it was naturally impossible to avoid any interaction with the 'mainstream' – however, it was defined – over the course of one's daily life. On the other hand is the fact that this blurriness was, in fact, an integral part of the countercultural experience itself. Moving to rural Wales was a way of putting the countercultural lifestyle into practice, but it was something that could not happen without interacting with Welsh culture – it was simply impossible to enact that side of the counterculture without also coming into contact with the Welsh 'mainstream'.

Both of these ways of blurring are integral to the stories shared here. The idea of the messiness of ordinary lives is embedded in virtually every contribution. The degree to which they interacted with elements of the 'mainstream' and 'counterculture' varied greatly – some were heavily involved with alternative lifestyles and only interacted with the 'mainstream' where absolutely necessary, others cheerfully mixed generous amounts of 'mainstream' in with their attempts at alternative living. This wasn't necessarily something that happened subconsciously, as you might expect – plenty of migrants were aware of the faint sense of impossibility that surrounded the idea of completely abandoning the mainstream. As Gillian and Freya commented:

Gillian: You felt that if you weren't reliant on the grid and you weren't reliant on the water supply you've got it all yourself.

Freya: Got your own stuff.

Gillian: You'd be able to survive. You knew you wouldn't but, but there was that feeling that . . .

Freya: You could always go and cut yourself peat and cut some trees and you, you could live your life without too much input from outside. Which again, which was tot-, we knew it wasn't real because we had a vehicle that used petrol, you know. And we'd never be able to produce enough of anything to, to pay for those kinds of things so, but it was just what you could do, you did do really.

I could, of course, have got a more homogenous story if I had been stricter about my definition of 'counterculture' when recruiting migrants. Rather than allowing them to self-identify and accepting anyone who moved in the right timeframe, I could have looked for people who shared specific, 'countercultural' characteristics. This might have lessened the sense of blurring between 'mainstream' and 'counterculture', giving a more unified image of a distinct movement. But realistically, bearing in mind what we have learnt, what characteristics could I possibly have chosen that would have done justice to the counterculture itself? Had I restricted it to smallholders we would have missed out on the significance of the entrepreneurs, had I focused on squatters we wouldn't have learnt about the wider cultural significance of DIY, and so on. Narrowing the definition of 'counterculture' would have produced a more coherent image, but it would not have been more accurate. It would have been in danger of playing into the hippy stereotype that I have been so keen to avoid. And it would not have revealed the limitless diversity of experiences in the lives of this group of migrants. This research would suggest that really the only way the counterculture, in its most all-encompassing sense, can ever be understood is on an individual basis – there are simply too many variables otherwise.

Besides, as the second point regarding why the boundaries between 'mainstream' and 'counterculture' were quite so blurry shows, even a regulated and homogenised definition of 'counterculture' would still have been subjected to blurred boundaries. As we know, it was simply impossible for the counterculture to involve relocating to rural Wales (or any other area for that matter) and not also involve interactions with the dominant local culture (not to mention the borrowing of certain aspects of rural culture discussed earlier). It has to be remembered, too, that the 'mainstream' that counterculture was blurred into was by no means a homogenous entity either. We saw in the final chapter how Welsh culture was regarded as something that was impossible to pin down, encompassing a far wider range of cultural components than appreciated by those outside Wales. There may have been elements of truth in the image of the 'rural idyll', but it was predominately as much a myth as the idea of a united, completely separate 'counterculture'.

You may be wondering what, if the ideas of 'mainstream' and 'counterculture' are so apparently meaningless, is the point of using them at all? Why not write off the counterculture as a myth, something which never really existed in the first place? However, the stories shared here have shown that although the line between 'mainstream' and 'counterculture' may be blurred, there is definitively a distinction there, somewhere. It may be impossible to isolate and dependent on context and individual circumstances, but these stories clearly show a belief in something, even alongside the awareness that it was impossible to be 100% immersed in it. As Freya said in the aforementioned quote: 'what you could do, you did do'. At the very beginning, I pointed out that the divisions often made between 'counter-urban migration', 'internal migration' and 'lifestyle migration' are rather arbitrary, as migrants almost always fit into multiple categories simultaneously. However, this does not stop the categories from being useful ways of thinking about the different forms migration can take and the implications of these differences. In the same way, the stories here show that while a division between 'mainstream' and 'counterculture' may be considered arbitrary given the considerable possible overlaps between the two, this doesn't stop the two concepts from being useful ways of thinking about different lifestyles, and the practicalities and implications of each.

The downside of the self-identification approach taken here is that this research does not cover any stories of extreme, die-hard countercultural experiences. It must be remembered that it is not, and was never intended to be, a representative sample of the counterculture. I have shown that the division between 'counterculture' and 'mainstream' was not at all clear-cut in practice, and people frequently blended aspects of the two. However, it is not possible to prove, based on the research contained in this volume, that nobody was committed to the counterculture on an extreme level, and hence, we must remember that these stories have not formed part of this book. I have shared the story of a part of the counterculture which has received little recognition before and which helps to explain exactly how the counterculture as a whole operated and fitted together (or not). However, it is not the story of the counterculture as a whole and nor does it claim to be. As I mentioned at the start, there are aspects of the counterculture which one might expect to be discussed which do not appear here – either because of ethical considerations or because they simply did not form a big part of the interviews – and it is important to remember that this does not mean that they were not a part of the counterculture at all. I have shared one set of stories, but there are still many others to be explored.

There are grand statements that could be made about the lessons to be learned here, sweeping speeches about the wisdom of the past and so forth. However, those are not the statements I want to make. It was never my intention to tell a generalisable story about the experiences of 'all' migrants. In addition, although time, space, and stylistic constraints have meant that I have had to group together certain experiences and highlight recurring themes, the ultimate beauty in this project has been in uncovering individual lives. The aim of the methodology was to prioritise ordinary voices, allowing everyday experiences to be heard. I wanted to draw on the potential of oral research to disrupt the dominant narratives, shifting the focus away from big events and major figures towards the ordinary and everyday. I did not expect that the story would end up being so much a story of reactions to those same big events that I wanted to divert attention from. Nor did I anticipate that these reactions would be quite as diverse as they were. I could never have hoped to find a way of generalising them all, even if I would wanted to. It has just as well that I wanted to focus on the individuals, the personal narratives.

To close, I want to take a moment to reflect on this diversity. Towards the end of Chapter 9, I noted that many contributors to this project would refer to themselves in contrast to the 'hippies on the hill'. This seemed to be the term used to refer to another set of countercultural migrants who were more completely immersed in an alternative lifestyle. For the contributors, this was a way of framing their identity and experiences, a way of saying that not only they were part of the movement but also they weren't as much a part of it as some people were. The trouble is that I never actually managed to find anyone claiming to be one of these 'hippies on the hill'. Nobody ever took this position at the top of the countercultural hierarchy. Maybe I just never found them. Or maybe, as I strongly suspect, there never were any 'hippies on the hill'. Maybe they were always a myth.

This project has shown that what seems on the surface to be a straightforward piece of a coherent countercultural movement is in fact far from being either straightforward or coherent. The counterculture was a tumultuous array of ideas and practices, collected together by a single heading. Judging by the stories shared here, I do not believe that it ever truly existed all in one place. The values and ideals of the counterculture dealt with deeply personal issues. They had the potential to impact every aspect of day-to-day life, from what you ate to where you lived, from your religion to your relationship with the natural world. The decisions that surrounded whether and how to put each potential aspect of the counterculture into practice were therefore also deeply personal. Of course, the countercultural migrants to Wales blurred the boundaries, borrowing from here and there, reacting all the time to what was happening around them. They were all people, after all, each with their own hopes and dreams and aspirations. The thing which unites them is that somehow, despite their infinite variety, they all made the same decision at the same time. At some point in 15 years, they all moved to Wales – they all went through the shared experience of deciding to go, finding a home, adjusting to a new life, and making a space for themselves in a new community. How they did these things was up to them.

References

Allaby, Michael and Tudge, Colin (1977) *Home Farm: Complete Food Self-Sufficiency*. London: Macmillan.

Chakraborti, Neil (2010) 'Beyond "Passive Apartheid"? Developing Policy and Research Agendas on Rural Racism in Britain'. *Journal of Ethnic and Migration Studies*, 36(3): 501–517.

Cherrington, John (1983) *A Farming Year*. London: Hodder and Stoughton.

Cloke, Paul, Goodwin, Mark and Milbourne, Paul (1998) 'Inside Looking Out; Outside Looking In: Different Experiences of Cultural Competence in Rural Lifestyles'. In Boyle, Paul and Halfacree, Keith (eds.) *Migration into Rural Areas: Theories & Issues*. Chichester: Wiley. Pp. 134–150.

Day, Graham (1998) 'A Community of Communities? Similarity and Difference in Welsh Rural Community Studies'. *The Economic and Social Review*, 22(3): 233–257.

Gemie, Sharif and Ireland, Brian (2017) *The Hippie Trail: A History*. Manchester: Manchester University Press.

Halfacree, Keith (2009) '"Glow Worms Show the Path We Have to Tread": The Counterurbanisation of Vashti Bunyan'. *Social & Cultural Geography*, 10, 771–789.

Halfacree, Keith (2011) '"Alternative" Communities in Rural Wales'. In Milbourne, Paul (ed.) *Rural Wales in the Twenty First Century: Society, Economy and Environment*. Cardiff: University of Wales Press. Pp. 65–88.

Nelson, Elizabeth (1989) *The British Counter-Culture, 1966–73: A Study of the Underground Press*. Basingstoke: Macmillan.

Robinson, Vaughan (1993) 'Making Waves? The Contribution of Ethnic Minorities to Local Demography'. In Champion, Tony (ed.) *Population Matters: The Local Dimension*. London: Paul Chapman Publishing. Pp. 150–170.

Roszak, Theodore (1969) *The Making of a Counter Culture: Reflections on the Technocratic Society and Its Youthful Opposition*. Berkley; Los Angeles: University of California Press.

Smith, D. M. (1999) 'Geography, Community, and Morality'. *Environment and Planning A*, 31: 19–35.

Tönnies, Ferdinand (2012 [1887]) 'Community and Society'. In Lin, Jan and Mele, Christopher (eds.) *The Urban Sociology Reader*. Abingdon: Routledge.

West, Elizabeth (1977) *Hovel in the Hills: An Account of 'The Simple Life'*. London: Faber and Faber.

Williams, Charlotte (2007) 'Revisiting the Rural/Race Debates: A View from the Welsh Countryside'. *Ethnic and Racial Studies*, 30(5): 741–765.

Yinger, J. Milton (1982) *Countercultures: The Promise and Peril of a World Turned Upside Down*. London: Collier Macmillan Publishers.

Index

Printed in the United States
by Baker & Taylor Publisher Services